NEW DIMENSIONS IN AGRICULTURAL GEOGRAPHY

VOLUME THREE

SOCIO-ECONOMIC DIMENSIONS OF AGRICULTURE

Concept's International Series in Geography

Volume 1 **Historical Dimensions of Agriculture**
Volume 2 **The Ecology of Agricultural System**
Volume 3 **Socio-economic Dimensions of Agriculture**
Volume 4 **Landuse and Agricultural Planning**
Volume 5 **Spatial Dimensions of Agriculture**
Volume 6 **Human Perception and Technological Change in Agriculture**
Volume 7 **Dynamics of Agricultural Development**
Volume 8 **Anthropogenic Dimensions in Agriculture**

Concept's International Series in Geography No. 4

NEW DIMENSIONS IN AGRICULTURAL GEOGRAPHY

VOLUME THREE

SOCIO-ECONOMIC DIMENSIONS OF AGRICULTURE

NOOR MOHAMMAD
Editor

CONCEPT PUBLISHIING COMPANY, NEW DELHI 110059

ISBN 81-7022-400-4 (Set)
ISBN 81-7022-403-9 (Vol. 3)

First Published 1992

Published by
Ashok Kumar Mittal
Concept Publishing Company
A/15-16, Commercial Block, Mohan Garden
NEW DELHI 110059 (India)

Lasertypeset by
Microtech Advance Printing Systems (Pvt.) Ltd.
H-13, Bali Nagar,
NEW DELHI 110015

Printed at
Sunil Printers
CB 386, Ring Road, Naraina
NEW DELHI 110028

PREFACE

AGRICULTURE CONTINUES to be the key industry. It is not only the dominant form of employment but also forms the basis for a way of life for, perhaps, two-thirds of the world's households. Agriculture being the function of physical, socio-institutional and techno-economic factors which are dynamic in nature, keeps on changing with the basic objective of increasing production and generation of foodgrains surplus. The traditional system of farming is in the process of transformation to modern agriculture which itself is undergoing a set of equally and generally even in more rapid changes. So it has added a new dimension in agriculture which needs to be studied and analysed.

A project was, therefore, taken up to accomplish this task. Original research papers dealing with various facets of agriculture were invited from scholars of different discipline from all over the world. The response was very encouraging and a large number of original papers were received. Besides these, some papers which have classical and everlasting value have also been picked up from my earlier publication – *Perspectives in Agricultural Geography*. The papers have been classified according to their basic themes and have been put into following eight volumes so as to give a systematic picture of various themes. The articles dealing with concepts, theories and models have been kept first followed by those dealing with the same theme at macro, meso and micro levels respectively.

The First Volume – *Historical Dimensions of Agriculture* – containing 19 articles deals with the origin, diffusion and development of agriculture, research trends in agricultural geography, world agricultural systems and methods and techniques in agricultural geography.

The Second Volume – *The Ecology of Agricultural System* – deals with the selected parameters of physical environment specially the climatic elements which affect the agricultural landscape in spite of technological advancement in agriculture.

The Third Volume – *Socio-economic Dimensions of Agriculture* –

containing 19 articles deals with socio-economic environment which have bearings on agricultural systems. The first 8 articles deal with social parameters specially labour, caste, women etc., while the last 11 articles examine the impact of techno-economic factors specially the various facets of irrigation which is a basic agricultural input.

The Fourth Volume – *Landuse and Agricultural Planning* – contains 19 articles dealing with concepts, theories, models of landuse, land capability and suitability, wastelands and dynamics of landuse.

The Fifth Volume – *Spatial Dimensions of Agriculture* – includes articles on agricultural systems of specific areas such as Himalayas, Bastar, Uttara Khand etc. It also deals with dynamics of crop landuse, cropping pattern and its diversification, crop ecology and crop-combination analysis.

The Sixth Volume – *Human Perception and Technological Change in Agriculture* – contains only 15 articles and covers four basic themes. The first set of 3 articles deals with human perception and his consequential response, while the second set includes articles on correlates of technological change and adoption of agricultural innovations. The third set of articles deals with the assessment of socio-economic transformation in rural India. The last set deals with agriculture and human settlements.

The Seventh Volume – *Dynamics of Agricultural Development* – includes articles on concept of development and its various facets. Besides assessing the trends of levels of agricultural development, it also examines and evaluates its impact on environment and ecological parameters.

The Eighth Volume – *Anthropogenic Dimensions of Agriculture* – containing 17 articles deals with regional imbalances in agricultural productivity, methods of its measurement, and also its correlates and determinant. It also focuses on food availability and food security at regional and continental level.

The collection of papers in eight volumes of *New Dimensions in Agricultural Geography* is designed to encapsule the major themes and recent trends and concepts in the field of agricultural geography with particular reference to conceptual revolution in the discipline. All the volumes contain a brief introduction which gives a glimpse into the topics dealt with in each of them.

The variety of articles contributed by the renowned scholars of different disciplines has made the character of these volumes multi-disciplinary and has enhanced their value. These volumes have been designed to help not only the students of a particular discipline but also to general students who wish to keep abreast of the latest developments in the field of agricultural geography. From these points of view these volumes hold a promise of rich fare and in addition to being heuristic, will provide a mine of information for discerning scholars. It is hoped that this unique collection of research papers will not only be valuable to the researchers in the fields of agricultural geography, economics, sociology, history, and agricultural science, but will also provide guidelines to decision-makers and planners.

The completion of this project has been possible due to the constant inspiration and unstinting co-operation of a host of people too numerous to acknowledge individually. However, I owe a debt of gratitude to my mentor — Dr. Mohammad Shafi, Professor Emeritus, Department of Geography, A.M.U., Aligarh who gave me valuable comments and suggestions ever since the inception of this project. Prof. R. Ramachandran and my other colleagues who offered sincere comments on the title of volumes and the collection of articles deserve special thanks. I would also like to thank my father, wife and children who provided a favourable working atmosphere to me for successful completion of this project. Last but not the least, I would like to put a special word of appreciation, love and affection to Mohammad Imran Ali Khan who is a source of hope and inspiration.

NOOR MOHAMMAD

Department of Geography
Delhi School of Economics
University of Delhi
Delhi-110007
January 26, 1992

PUBLISHER'S NOTE

When Concept's International Series in Geography No. 1 — *Perspectives in Agricultural Geography* edited by Dr. Noor Mohammad in five volumes was issued over a decade back, we expressed our apprehensions that 'Posterity will be the best judge of our efforts in this direction. The response to this set of volumes will provide the impetus for hastening the publication of further series that are already on the anvil'.

Perspectives in Agricultural Geography received such an overwhelming response that it became out of print soon after its publication. Its success provided the necessary encouragement to issue thereafter Concept's International Series in Geography No. 2 — *Perspectives in Urban Geography* edited by Dr. C.S. Yadav in 16 volumes and No. 3 — *Perspectives in Geomorphology* by Dr. H.S. Sharma in four volumes.

We are glad to offer now Concept's International Series in Geography No. 4 — *New Dimensions in Agricultural Geography* edited by Dr. Noor Mohammad in eight volumes. In view of the new emerging trends, concept, theories and models in the field of agricultural geography, the set presents a wide panoramic view of the latest advances in this field. Hope this set too would be equally welcomed by the students, researchers, scholars and planners in various disciplines.

February 11, 1992 NAURANG RAI

CONTENTS

LIST OF FIGURES

LIST OF TABLES

LIST OF CONTRIBUTORS

Aparna, S.

Formerly Research Scholar, Centre for the Study of Regional Development, Jawaharlal Nehru University, New Delhi

Bharara, L. P.

Social and Information Science Division, Central Arid Zone Research Institute, Jodhpur

Coutinho, O.

Department of Geography, Rani Parvati Devi College, Belgaum

Das, Debashis

Department of Geography, Visva-Bharati University, Santiniketan, Birbhum

Gregor, Howard F.

Department of Geography, University of California, Davis, U.S.A

Ilbery, Brian W.

Department of Geography, Coventry Polytechnic, Priory Street, Coventry CV15FB (UK)

Jain, C. K.

Department of Geography, H.S. Gour University, Sagar

McKnight, Tom

Department of Geography, University of California, Los Angles, California, USA

Mohammad, Noor

Department of Geography, Delhi School of Economics, University of Delhi, Delhi

Qureshi, M. H.

Centre for the Study of Regional Development, Jawaharlal Nehru University, New Delhi

Ram, Balak

Central Arid Zone Research Institute, Jodhpur

Reddy, K. Mohan

Department of Economics, Kakatiya University, Warangal

Reddy, K. Sateesh
Department of Economics, A.P. Open University, Hyderabad

Reddy, M. V.
Department of Geography, S.V. University, Tirupati

Reddy, N. B. K.
Professor of Geography (Retd.), Department of Geography, S.V. University, Tirupati

Sen, A. K.
Central Arid Zone Research Institute, Jodhpur

Sharma, S. K.
Department of Geography, Dr. H.S. Gour University, Sagar

Sharma, T. C.
Department of Geography, Rani Parvati Devi College, Belgaum

Singh, Lakhvinder
Department of Geography, Lyallpur Khalsa College, Jalandhar

Sinha, B. R. K.
Development of Geography, Visva-Bharati University, Santiniketan

Stola, Wladyslawa
Institute of Geography and Spatial Organization, Polish Academy of Sciences, Warsaw, Poland

Varma, Shashi Kanta
Department of Home Science Extension Education, College of Home Science, Haryana Agricultural University, Hisar

Wilken, Gene C.
Professor of Geography, Department of Economics, Colorado State University, Fort Collins, Colorado, U.S.A

INTRODUCTION

THIS VOLUME containing 19 research papers deals with socio-economic environment which have a strong bearing on agricultural system. The geographers, in general, have used to emphasise the role of physical environment, whereas economists have laid more emphasis on economic factors in conditioning the distribution of geographic phenomena specially the economic activities, and as such the natural relationship between the phenomena has been under emphasised (Morgan and Munton, 1971 and Chisolm, 1966). In fact farming is the product of not only physical environment but also of social and technological framework such as land ownership and tenancy system, size, shape and layout of farms, pressures of population and farm workers densities, the ensemble of farm buildings, agricultural implements, irrigation system, location of market centres, attitude of farmer and his literary status (Singh, 1974). Hence, launching a direct attack on technological problems alone will not be sufficient to ensure success for agricultural development plans and package programmes, but simultaneous attention must be given to the social institutions which at the moment are impediment to the rapid development of agriculture. So much intimate is the relationship between agriculture and character of social institutions and customs, that it becomes virtually impossible to decide which is cause and which is effect (Buchanan, 1959). However, in recent studies attempts have been made to analyse the agricultural systems in this light (Chisolm, 1966; Myrdal, 1971; Hodder, 1971; and Logan, 1972).

There is no general theory of social relations which could be fairly applied to explain spatial distribution of agricultural systems, but social differences within communities and between communities affect scale and type of farming operations together with choice of enterprise. In view of world differences of farming, it is hardly surprising that agricultural operations should be the subject of social constraints and taboos and that land should be set apart from other forms of property, be vested in the community or even be regarded as object of religious worship. The social systems sets its imprint on holding and field systems and on the settlement pattern with its related problems of accessibility to field (Morgan and Munton, 1971).

Similarly the techno-economic factors also play significant role in not only shaping the agriculture but also in its development. The technological factors throughout the history of mankind in general and during the recent past in particular have played a significant role not only in overcoming various environmental constraints on agriculture but also in bringing further changes and development in existing ones. The most obvious example is the introduction of canal system of irrigation in arid region of Rajasthan (India) which has replaced the extensive pastoralism with intensive cultivation; or the example of sandy soil in Netherlands, which has been so enriched and co-settled over the years that it produces the world famous bulb crop from an artificial soil could also be cited (Paterson,1976). Similarly in order to control the adverse climatic conditions or rather to offset the climatic extremes, technological system at certain socio-economic level have given rise to such varied means as irrigation, greenhouse, smudge pots, afforestation and cloud seeding. Methods have been developed to increase productivity by using fertilisers and better techniques of working the soil. Control of life process of plants and animals has been achieved through improved breeding, development of hybrid researches into such basic processes as photosynthesis, and finally control of pests and weeds through the development and manufacturing of pesticides, fungicides and herbicides (Hurst, 1972).

In brief the technological change in agriculture consists of adoption of farming techniques developed through research and calculated to bring out diversification and increase in production and greater economic return to farmers. The use of fertilisers, improved varieties of seeds, pesticides and fungicides, improved irrigation facilities, new farm implements and contour bunding for the conservation of moisture and soil are some of the examples of such techniques (Mohammad, 1981). However, it is not possible to include articles on all these parametres in this section but certain aspects have been covered in detail at macro, meso and micro levels.

Although different methods are used to measure the agricultural productivity in different countries depending on their socio-economic condition, it is, by and large, expressed in terms of per unit of land and labour. The common finding of almost all the studies on this theme is that as compared to traditional agriculture, modern farming has much high rate of productivity per unit of labour and land. **Wilken** in his paper argues that the comparison of two systems on different levels of integration is unfair. Such comparisons are fallacious because they treat men and farms as comparable unit rather than as functional and spatial elements included in fundamentally different systems. The author with various concrete and specific examples has highlighted that in traditional agriculture the total input in the form of land and labour is almost direct and measurable, whereas in modern agriculture, it is not so because it comes through various direct and indirect sources. Since the inputs are not measurable, it is difficult to say that modern agriculture is more productive than traditional one, although the author is hesitant to make this statement.

Caste and primary occupations are so closely linked that they are quite often taken as synonym to each other. In the past every caste had an occupation of its own which used to be hereditary and to abandon it in search of others was considered to be sinful. But with the passage of time both have changed considerably. The purity of caste exists only in religious books. Occupational structure has also undergone a sea change due to technological advancement. However, under such an overall caste-occupation dynamism, caste-occupation relationship still exists at micro level. **Mohammad** in his paper has assessed the impact of caste on primary occupation in a predominantly agricultural state of Uttar Pradesh. The study is based on primary data collected from 1033 respondents chosen from 35 villages located in different caste-combination regions. The author has selected 12 major castes and six occupations — crop farming, dairy farming, vegetable farming, fishing, poultry farming, piggeries, service and labour — in addition to other occupations and has assessed the relationship between the two. The analysis reveals that the degree of relationship depends, by and large, on level of technological advancement and modernisation. In the backward isolated areas the primary occupations are still related with certain castes such as crop farming with Brahmin, Rajput, Jat and Kurmi, dairy farms with Ahir and Jat, poultry farming with Sheikh and Scheduled Castes, piggeries with Scheduled Caste and Lodh, and vegetable farming with Kori and Kachhi. Since it is their hereditary occupation, they know all its techniques, methods ins and outs leading to higher production. We should, therefore, encourage the development of primary occupation according to the caste which have it on hereditary ground and micro level planning should be done on this basis.

Human resource is considered as an active agent of production which does not accumulate and exploit only nature's bountiful resources and increases the productive potential of a region but also enhances and build social, economic, cultural, political strength and ultimately carry forward national development. It helps in the process of production of usable and consumable social values so that entire mass of the people can live a life under condition of reasonable comfort after satisfying their basic needs. **Sinha** in his paper has made a maiden attempt to assess and analyse the human resources in Dhanaura block of Patna district in Magadh Plain of Bihar. The study is based on primary data collected from 20 per cent households of 14 selected villages – 8 Cultivator Dominated (CD) and 6 Agricultural Labour Dominated (ALD). The study impinges on 5 to 23 years age groups of student population which is a critical age of human resource formation. The author has made a comparative analysis at age and sex composition at different levels of education in the CD villages and ALD villages and also at block level which indicates a lot of variations at almost all levels of analysis. The author has correlated the human resource with rural development and has finally suggested various measures for proper development of the human resource.

The prosperity and growth of a nation depend on the status and development of its women as they not only constitute nearly half of its population, consume two-thirds of the world's working hours, earn one-tenth of income and own one per cent of world property but also influence the growth of the remaining half of the population. Although women contribute in various sectors of economy but it is more so in agriculture. **Varma** in her paper: (a) assesses the nature and extent of involvement of rural women and men of various socio-economic strata in farm, household and dairy activities in terms of operations, time spent, responsibility and accountability; (b) examines the comparative drudgery of men and women for agricultural operations and relationship with socio-economic correlates, and (c) assesses the women's exposure to modern crop technology and role of women in decision-making in farming, house and other activities. The study concludes that men are not callous to women and they do not consciously put women to hardship by assigning them the task of greater drudgery. In fact men and women both are unfortunate victims of some unrealistic, unreasonable and undesirable social values. New technology coupled with a well designed programmes of education for rural men and women will help to achieve quality of life.

Agriculture is the kingpin of the tribal economy. The nature of tribal economy being dichotomous, i.e., agriculture and forestry farming its base, a significant percentage of the working force is engaged in these two sectors. Nearly 80 per cent of the tribals live in rural areas and 93 per cent of them depend on agricultural and allied sectors. **Reddy** and **Reddy** have examined the absorption of human labour — tribal and non-tribal at different levels of agricultural operations in the cultivation of three major crops viz. paddy, jowar and gingelly in Warangal district of Andhra Pradesh. The study is based on primary data pertaining to labour absorption collected from 153 households selected from 9 villages following three Mandals viz. Eturnagaram, Tadvai, and Kothagudem. From each village 10 tribal households and 7 non-tribal households have been selected. The authors have identified 5 main agricultural operations — ploughing, sowing, weeding, harvesting and threshing together with other operations, and have tried to analyse the labour absorption in terms of male, female, owned and hired labours among the tribal and non-tribal population in case of paddy, jowar and gingelly separately. The study concludes that the absorption of hired labour is more than that of owned labour for the cultivation of paddy, whereas it is marginal in case of jowar and gingelly. The absorption of female labour is also high in case of paddy cultivation when compared to jowar and gingelly. The absorption of tribal labour in absolute term is more for growing jowar and gingelly, whereas it is less absorbed for cultivation of paddy when compared to non-tribal labour force.

Stola in his paper has made a functional classification of communes in Poland on the basis of 10 attributes, which are related to agriculture, forestry, recreation and non-farming functions such industry, social amenities etc. The study has distinguished seven principal classes of communes differing from one and other in their leading functions. These are highly urbanised and semi-urbanised, industrial agricultural, non-sociolised agricultural, tourist and recreational, forestry and agricultural communes. In fact, this classification is basically done on the basis of the dominance of the function of different communes. The study reveals a lot of spatial differentiation in the functions of various communes which shows a great need for individual approach while making decisions for its spatial planning.

In the Third World land tenure system is very varied and complex and is major constraint in agricultural development programmes

(Morgan, 1978). The problem is more intricate in India. Prior to Independence, the British introduced various land reform systems which were obviously geared towards better collection of revenue. But in the middle of 19th century, three basic land revenue systems have been gradually introduced in India viz. *zamindari, ryatwari* and *mahalwari*. In *zamindari* system the intermediaries form an important segment in the chain of relationship, whereas *ryatwari* brought the state and the tiller in direct relationship, while *mahalwari* is a system envisaging communal collection of land revenue. Since Independence, the centre and the state governments are introducing various land reform acts for better utilisation of resources and removing the income disparities in the rural areas. The success in this direction depends not in the enactment of a host of acts and laws but in their implementation. We are aware of the fact that no legal provision is perfect and free from loopholes under the existing social set-up. **Qureshi** and **Aparna** in their article have examined critically various land reform measures introduced in the state of Kerala and their impact on land tenure systems. The authors have given an overview of land reform measures in the three erstwhile units of the state viz. Travancore, Cochin and Malabar at various points of time before the formation of the state in 1956. Finally they have examined various land reform acts introduced in Kerala after 1956 and their impact on overall distribution of land.

In fact, the impact of land reform measures is reflected in the pattern of land distribution and can be roughly measured by looking at the concentration of land in various size class of holdings. The authors have done this exercise also by applying Ginni's co-efficient separately for total holdings, wholly owned and self operated holdings, and partly owned and partly rented holdings. Lorenz curves have been drawn for individual districts and state as a whole in order to show the spatial pattern which is inclined toward concentration of land in the small size classes.

With the modernisation and technological development which is percolating gradually in rural areas, occupational structure is becoming complex day by day. People do not get satisfied with a single occupation and in several cases it does not fulfil the necessities of the household as well. It is more so when the family is large and the demands are varied. Therefore, not only the head of the household but also the members of the family go for several subsidiary occupations which contribute to the family income. On the basis of the importance – social and financial, the occupations may be divided in hierarchical order which would be helpful in planning overall socio-economic development of the region. The hierarchy of occupation varies from one caste to the other caste. **Mohammad** in his paper has described and explained the caste and hierarchy of occupations in a predominantly agricultural state of Uttar Pradesh. The study is based on primary data collected from 1033 sample respondents chosen from 35 villages falling in different caste combination regions. The author has taken six main occupations — crop farming, vegetable farming, dairy farming, poultry farming, fishing and piggeries, and 12 major castes. He has also worked out the hierarchy of occupations in terms of first, second and third occupation of the selected occupations of different castes of the state. The study comes up with 12 significant conclusions which may be taken as guideline for planning overall development of the state specially the agricultural development.

Capital has always occupied a place of significance in the farming system and it is being now more recognised as a growing worldwide phenomenon with the process of mechanisation of farming. In the industrialised countries it has brought huge production gain but has also created social and resource problems whose full seriousness is only beginning to be realised in the more agrarian countries. It is regarded as absolutely essential, although there is a debate over the forms and depth of capitalisation to be sought. Yet despite this prominence, little attention has been given to agricultural capitalisation in a spatial way, and that has been presented mostly in general world regional terms (Herlimann, 1954). **Gregor** has made a detailed and spatial analysis of agricultural capitalisation in U.S.A. with particular reference to southwestern corner, where some of the highest as well as the lowest levels of capitalisation are attained. High order factor analysis is applied which reveals that capitalisation is widely distributed but its most important factors – cropping emphasis, intensive landuse, and managerial efficiency, favour the better farmland and the areas closer to cities. With amalgamation of the factors, urban influence proves to be the dominant spatial pattern of capitalisation despite the fact that the most important production areas do not locate nearest the cities and loss of agricultural land on urban margins continues to be high.

The importance of irrigation in the development of agriculture in general and in arid and semi-arid regions in particular hardly needs any emphasis. The vagaries of monsoon has made it imperative for India to accord irrigation a high place in the national development plans.

Among all Indian irrigation systems, minor irrigation system has made comparatively more substantial progress specially in the arid region. It has received further special emphasis with the inception of Drought Prone Area Programme. However, minor irrigation system is very much embedded into the social systems, values and norms. Different socio-economic and cultural factors such as kinship, caste based joint ownership of land and water resources, norms towards water use etc. characterise its solidarity and continuance since generations. But with the passage of time the system has lost much of its social significance. Various impediments occurring in the system inhibit the irrigated cultivation, consequently socio-economic development in the area. In order to overcome it, there is a need for design of a new or modification of the existing social norms of the system which will ensure an effective combination of resources and people in terms of land-water relationship. **Bharara** in his paper has discussed this problem at micro level. The study was conducted in an arid village — Daijar. All sources of irrigation were studied and supplemented with field data collected in 1979 through personal interviews of the heads of the households having irrigation sources at different scattered farms of the village. Representatives of various caste groups were interviewed. The author has studied the type of minor irrigation system and its socio-economic characteristics, source-wise irrigation potential available, ownership coverage and distribution pattern, control of water use, cultivations and irrigation turns, and the resultant policy implications to safeguard against irrigation insecurity due to drought. The author has made certain suggestions to overcome these problems.

In the post-war period many western countries have undergone second agricultural revolution characterised by concentration of capital and output on fewer larger farm business and an increasing specialisation and intensification of production resulting not only to self-sufficiency in food production but generating adequate surplus production. This process through huge production – demand and price mechanism has resulted into indebtedness and low income among the farmers of the many western countries. It has, therefore, resulted into restructuring of farm business in western countries. **Ilbery** in his paper has examined this issue and has suggested alternatives.

Although various possibilities have been suggested to overcome this issue, the author has examined farm diversification as most appropriate alternative enterprise. Various explanation and strategies are suggested for farm diversification but the most encompassing definition is "Those enterprises taking place on agricultural proprietal unit which: (a) are not based on primary production of food and fibre and/or (b) fall outside the price support mechanism of common agricultural price (Slee, 1977)". It gives rise to five main group of activities which are tourism, recreation, adding value to conventional products, unconventional enterprises and use of ancillary buildings and resources. The author examining the factors affecting farm diversification has described mechanism and the development of farm diversification on urban fringe. The problem involved in the development of farm diversification has also been discussed by the author. The study concludes that the future for farming in western world is uncertain yet it is certain that major changes in agricultural landuse will take place before the end of the 20th century.

Water is one of the most essential inputs for agricultural development but its availability is conditioned by a number of physical, social and techno-economic factors. Hence, there is spatio-temporal variation in irrigation facilities. **Sharma** and **Jain** have examined the impact of irrigation on agricultural development in Madhya Pradesh. After assessing the present status of irrigation facilities, they have evaluated the progress of source-wise irrigation during 1950-51 and 1976-77 which shows a phenomenal growth in it. Consequently the net sown area and total cropped area have also increased considerably during this period. Besides this, the impact of the rapid growth in the irrigated area on cropping pattern, adoption of HYV of seed, use of chemical fertilisers have also been examined and explained at district level which shows considerable spatio-temporal variation. There is positive relationship of varied level between irrigation and afore-mentioned parametres. However, this study concludes that Madhya Pradesh is agriculturally backward. Like other developmental measures, development of irrigation is a localised phenomenon. There seems to be no direct relationship between the amount of rainfall and intensity of irrigation. The percentage of irrigated area increased rapidly where its intensity was low and very low in the initial year, i.e., in 1950-51. The expansion of net area sown and reduction of cultivable wastes cannot be ascribed safely to the development of irrigation. However, irrigation has played a vital role in raising the yield rates of certain crops, such as rice and wheat, and horticultural crops. This is made possible through the adoption of certain yield-raising inputs, which are dependent, to a great extent, on irrigation. Use of the high-yielding variety of seeds and chemical fertilisers varies directly with the intensity of irrigated area, and have a significantly high positive correlation. The districts adopting these three measures, viz., irrigation, HYV of seeds and fertilisers also have a high agricultural productivity per unit of cropped area. It may be summed up that the development of agriculture in the state is tagged with the development of irrigation. Only the surface water resource is partially used for the purpose, leaving reserves of the underground water resource untapped. Therefore, use of the water resource may be planned accordingly.

Irrigation is not an early development in Australian agriculture but once it was initiated, not only its spread was rapid but high level irrigation technology has also developed in short duration of past four decades. In Australian context, high irrigation technology irrigation owes in three modes – lesser levelling of land for furrow irrigation, drip or tridel irrigation and complex automatic sprinkler irrigation. **McKnignt** in his paper describes the sprinkler irrigation system in Australia. Although sprinkler irrigation was introduced in the beginning of the 20th century, a lot of improvements have been made after World Wars due to the development of other infrastructures. The author has discussed in detail giving the merits, limitations and areas of application of Centre Pivot Technology and the Linear Move or Lateral Move. The author basically concentrates on application and distribution of automated sprinkler irrigation in Australia specially on Pivot system and Linear or Lateral Move. The study is based on primary data collected from 40 interviews and 105 circulated questionnaires of which only 57 per cent were responded but it was from all over the country. The author has not only given the spatial distribution but has described the associated cropping pattern with each system of sprinkler irrigation.

Irrigation is one of the basic inputs for agricultural development but its utilisation depends on a number of physical, social and technological factors. Among the socio-institutional factors the socio-economic relations and functional dependence among the farmers belonging to different caste groups is most prominent. Very often unhealthy functional relations among the farmers of different socio-economic groups lead to improper and unscientific use of available water resources leading to serious tussles among them. The strong group always gets on large share of the cake and it remains unchallenged. **Debashish Das** in his paper has examined this issue in a predominantly agricultural district of Bhirbhum of West Bengal. The district has surface and underground sources which are utilised through various systems of irrigation. It is inhabited by large number of high, medium and low caste people. The empirical study is based on primary data collected from 360 respondents selected on purposive sampling basis from sample villages. The author has described and examined with simple statistical techniques the inter-caste relationship between population belonging to different caste groups and irrigation, relationship between farmers belonging to different caste groups and net irrigated area, net value of crops and source-wise irrigated area. With a detailed statistical analysis of the primary data, the study summarises that the high and middle caste Hindu population have well control over all sources of irrigation, but scheduled caste farmers have large share of irrigated land that are irrigated by canal and river lift irrigation scheme. Scheduled tribe irrigate their land with river lift irrigation scheme, while Muslims have higher proportion of land irrigated by canal and shallow tubewells.

Irrigation is a decisive factor in Indian agriculture due to inadequacy and high variability of rainfall. It is more so in arid, semi-arid and sub-humid regions of the country which are prone to drought and famine conditions due to partial failure and delayed arrival or early withdrawal of the monsoon. Therefore, constant efforts are made to develop and improve irrigation system to mitigate this problem as much as possible. **Reddy** and **Reddy** in their paper have examined this issue in the state of Andhra Pradesh which consists of three major physical units — the Coastal Plain, the Eastern Ghat and the Interior Plateau, having different climatic conditions and water resource potentialities. The authors after accounting the irrigation potential of the state, have given the various sources of irrigation — canal, tank and well – for the year 1982-83. The changing pattern of irrigation is examined during 1972-73 and 1982-83 according to sources of irrigation and also in three regions of the state — Costal, Telangana and Rayalaseema – which shows source-wise a lot of spatial variations in it. The authors have done in depth study of irrigation pattern and type by adopting three different approaches, i.e., irrigation intensity, irrigation concentration by applying Florence location quotient index formula and irrigation orientation by applying Kostrowicki's formula in different regions of the state. However, the study concludes that as surface water resources are unequally distributed, the basin-wise planning for tapping the water resources may not solve the problem of the central and southern region. Maximum justice both to the region which has large water sources and to deficit parts of the state could be done, if the three river basins are integrated into one system through inter-basin transfer of water.

The state of Uttar Pradesh has made rapid progress in irrigation during the past three decades and about 53 per cent of the net sown area is under irrigation which is next to Punjab, Haryana and Tamil Nadu. But there is a lot intra-regional variations in it. Therefore, there is a need to formulate well conceived long term perspective irrigation development plan so as to utilise vast surface and groundwater resources to eliminate inter- and intra-regional variations in the levels of development. **Coutinho** and **Sharma** in their paper have examined disparities in the growth of irrigation — source-wise and in net and gross irrigated areas between and within the five agro-climatic regions of Uttar Pradesh, problems resulting due to haphazard and unplanned manners of irrigation facilities development, and measures necessary for a fuller exploitation of the state's potential surface and groundwater resources necessary for a balanced and faster growth of agricultural production. The study reveals that there is rapid expansion in the tubewell irrigation — a modern, scientific and efficient method of exploiting groundwater, contributing about 52 per cent of the state's net irrigated area. The canal irrigation is almost stagnant, while the contribution of other wells to irrigated area has declined. Tanks and other sources are of marginal significance and contribute only 5 per cent to net irrigated area. In spite of changing position of sources of irrigation, inter-regional contrast persists in terms of irrigated area and its development. The Western and Central regions of the state have more irrigation facilities as compared to eastern, southern and hill regions. Bundelkhand continues to remain the least developed region. The study concludes that U.P. is confronted with several problems due to lack and misuse of irrigation resulting from absence of a perspective planning for the development and management of its water resources.

Sharma and **Jain** in their article have examined the perspectives of irrigation in Madhya Pradesh with a hypothesis that the variation in proportion of irrigated area may be related with the variation in the need of irrigation, availability of irrigation water resources, and the capacity of the farmers to use it. The authors have described and explained the spatial pattern of irrigated area during 1985-86 which shows a lot of inter- and intra-regional variations in it. The source-wise progress of irrigation during 1950-51 and 1984-85 also shows almost similar patterns. A cross-classification of irrigated area and growth of irrigated area shows that more progress has taken place in the medium and low irrigated areas as compared to high irrigated area which has, in fact, recorded the lowest. In order to have indepth understanding of the irrigation system in the state, the authors have examined it in a variety of ways such as means of irrigation, rainfall and irrigation, land capability and irrigation, and socio-economic perspective and irrigation. Among the socio-economic perspectives literacy among cultivators, scheduled tribe population, size of holding, consciousness of the people and adoption of high raising inputs — high yielding varieties, and chemical fertilisers, have been taken. The study suggests that the development of minor and even micro-irrigation projects to utilise large water potential of the state for the development of agriculture, would be more economical and useful in the contemporary socio-economic conditions.

In the fragile ecosystem of the desert region water and its proper utilisation and management is the main problem of agricultural development. Recently a significant expansion of irrigation farming has taken place in Indian desert through exploitation of groundwater and construction of reservoir and canals which have resulted in salinity/sodicity, lowering of watertable, seepage and waterlogging which is causing concern to farmers, planners, scientists, and decision-makers alike. **Ram** and **Sen** in their paper have made an appraisal of present irrigation farming system, its change over time and space and scope for future development. Resultant problem areas have been identifical, delineated and mapped, and their impact on cropping pattern and crop production is assessed together with suitable suggestions for development of potential areas and judicious use and management of this valuable liquid. The study is based on remote sensing techniques and interpretation of survey of India topo-sheets supported by field survey and other available information. The authors, after describing the present landuse, have examined in detail irrigation farming with special reference to its present situation, cropping pattern, changes in irrigated area, changes in sources of irrigation, and changes in cropping pattern and crop production. The authors have also critically assessed the consequences of irrigated farming such as salinity/sodicity, lowering of groundwater, seepage and waterlogging and other problems, and have finally suggested suitable management practices. The study concludes that irrigated farming has vast potentialities but it needs wise and judicious use and management of both surface and groundwater supplemented with rational planning, political will and peoples' awareness.

Modern agricultural inputs specially chemical fertilisers have brought a major breakthrough in the development of agriculture but the application of heavy doses to recoup soil fertility results into serious detrimental consequences such as soil becomes desertic, barren and loses humus and other micro-organism. It also affects human health by contaminating water and food chain. **Singh** in his paper has assessed the use of chemical fertilisers and its implications, and has suggested natural farming in order to have sustainable development. The author has examined the used of chemical fertiliser in different crops and its impact which is positive on food production in Punjab, Haryana and Uttar Pradesh. Since the country has not only achieved self-sufficiency in foodgrains but also has buffer stock, it is high time, keeping the ill-effect of heavy doses of chemical fertiliser, to slowly and steadily switch over to organic manure and made a shift towards natural farming system. He has given a brief account of various organic manures such as farmyard manure, green manure, concentrated organic manure, bio-fertilisers and poultry waste, including their sources, characteristics, availability and use. Finally, he strongly pleads for the adoption of natural farming which is now essential for sustainable agricultural development.

REFERENCES

Board, C., et al., 1969 : *Progress in Geography*, Vol. I.

Buchanan, R. O., 1959 : 'Some Reflections on Agricultural Geography', *Geography*, Vol. 44, pp. 1-2.

Chisolm, M., 1966 : *Rural Settlement and Land Use.*

Herlimann, H. H., 1954 : *Techniesierungsstufen der Landwirstschaft Berchite Uber Landwirstchaft*, Vol. 32, p. 335.

Hodder, W. B., 1971 : 'Natural Resources and Tropical Development', in A. B. Mountijoy, (Ed) *Developing the Underdeveloped Countries.*

Hurst, M. E. E., 1972 : *A Geography of Economic Behaviour : An Introduction*, p. 83.

Logan, M. I., 1972 : 'The Development Process in the Less Developed Countries', *The Australian Geographer*, XII(2), pp. 146-153.

Mohammad, N., 1976 : "Technological Change and Diffusion of Agricultural Innovations", *The Geographer*, Vol. XII, Aligarh.

Mohammad, N., 1981 : *Perspectives in Agricultural Geography*, Vol. II, Concept Publications Co., Delhi.

Morgan, W. B., 1978 : *Agriculture in the Third World*, p. 117.

Morgan, W. B. and Munton, R. J. C., 1971 : *Nature of Agricultrual Geography*, p. 13

Myrdal, K. G., 1971 : *The Challenge of World Poverty : A World Anti-Poverty Programme in Outline.*

Norman, Long, 1977 : *An Introduction to the Sociology of Rural Development*, p. 145.

Paterson, J. H., 1976 : *Land, Work and Resources : An Introduction to Economic Geography*, p. 26.

Singh, J., 1974 : *An Agricultural Atlas of India : A Geographical Analysis*, Vishal Publication, Kurukshetra, p. 67.

Slee, B., 1987 : *Alternative Farm Enterprises*, Farm Process.

1

THE FALLACY OF LABOUR/LAND COMPARISONS OF AGRICULTURAL PRODUCTIVITY

GENE C. WILKEN

MOST DISCUSSIONS of agricultural development rest upon expressed or implied comparisons which indicate that the productivity per unit of labour or land is much lower in less-developed than in developed countries. In fact, the whole argument for modernisation of agriculture and transfer of technology follows from this premise. But such comparisons are fallacious since they treat men and farms as comparable units, rather than as functional and spatial elements imbedded in fundamentally different systems.

The characteristics of modernisation are well known: compared to traditional agriculture[1] modern farms are more capitalised, commercialised mechanised and specialised. They also are more integrated with markets and supporting industries.[2] In modern agricultural sectors farm sizes are large, populations are low, and overall productivity is high. The net result, derived by simple division, is that compared to traditional agriculture modern farming has much higher ratios of productivity per unit of labour (e.g., man-day) and land (e.g., hectare).

But can yields from systems at different levels of integration be compared? The situation is deceptive. In the equation: production÷(labour) (land)=yields, the crop numerators have a degree of universality that validates comparisons. Unlike manufacturing, where not only product quantity but also quality and even function change drastically with production methods, all agriculture utilises the same biological growth processes. Thus, although one would hesitate to compare the output of hand- with machine-made containers or shoes, entirely different farming systems can produce cotton or maize that is indistinguishable.

But the arithmetic used to develop the labour and land denominators differs markedly between systems. Labour in traditional farming is primarily direct and developed on-farm; in modern systems it is more managerial with labour inputs emanating from many parts of the economy. Similarly land in traditional agriculture usually encompasses little more than the farm itself whereas in modern systems the area occupied by a host of supporting activities contributes to farm production.

The situation can be illustrated with examples from opposite ends of the farming spectrum. Consider a traditional or non-industrial farmer operating with few labour, energy, or material inputs other than those from the farm area itself. Field preparation, cultivation and harvesting are performed by the farmer and his family, as are most support activities such as tool manufacture and repair, transportation, crop storage and training. While cases of complete self-sufficiency are rare (Clifton, 1969), the description is not far from the model often employed to characterise traditional or non-industrial farmers.

At the opposite pole, consider a modern or industrial farmer operating within a technologically and commercially integrated society. The farmer is less a direct unit of labour and more a manager of labour, energy and materials.[3] Disregarding on-farm use of labour-replacing fossil fuel itself, the efforts of countless miners, factory workers, engineers, servicemen and managers are represented in the machinery that transforms energy into power, and in other purchased inputs[4] such as fertilisers, improved seeds and irrigation water. Presumably services such as research, education, insurance and market analyses also increase efficiency and productivity. In effect, the modern farmer commands an army of workers, technicians, scientists, educators, financiers, administrators and analysts that figuratively labour beside him in his fields.[5]

These results of capitalisation or modernisation of agriculture are well known, as is the contribution of labour represented by the various inputs.[6] Yet statements of productivity continue to be based upon simple direct labour/land figures which deny the very essence of sectorial integration. One of the most common, for example, is that the five per cent of the United States population that is on-farm feeds the remaining 95 per cent.[7] Actually, the five per cent represents some 10 million managers controlling and benefiting from the labour of a vast number of people who work not only in farm-related industries but also in all activities that directly or indirectly affect farm productivity.[8] The question is not how much is produced by farmers, but how much is produced by labour inputs from the whole system. Dividing farm output only by direct farm labour ignores the traditional farmer's productivity in the various support activities, and the contribution of external support activities to the industrialised farm's productivity.[9] Failure to consider the labour and production systems within which each type operates renders comparisons invalid.

Theoretically all inputs in a completely commercialised society, including the managerial talents of farmers, are priced and farm productivity is sufficiently high to cover the costs. But not all costs are fully calculated nor charged against farm operations. For example, it is common to shift certain costs from specific producers or users to the general society. Disregarding direct subsidies for farm products and operations tax-supported services such as roads, schools, research and technical and administrative assistance make units of farm labour more efficient and productive but are only partially paid for by farmers.[10] To a certain extent these services are offered at low or no cost to farmers in all countries. But the level of services, and thus the degree of influence upon productivity is vastly greater in developed than in less-developed societies.

In addition, it is increasingly apparent that the costs of many material inputs are not fully calculated. This is especially true with respect to the manufacture and use of labour-replacing energy and equipment, in which environmental and other social costs are consistently deferred (Howard, 1971). Less-developed societies do not calculate these costs, or assign them to primary beneficiaries either. But energy use is much greater in developed economies and permeates almost all sectors. The effect is to establish a cumulative labour subsidy for the industrialised farmer and to put the non-industrialised farmer at an additional accounting disadvantage.[11]

The same type of reasoning applies to comparisons of productivity per unit of area. Contrasting yields per hectare produced by non-industrialised farmers with those of their industrialised counterparts ignores the spatial aspects of the integrated systems within which the latter operate. If the modern farm receives labour inputs from miners, factory workers, scientists and administrators, its productivity depends in part upon the space allocated to their activities. Appropriate shares of open-pit mines, assembly lines, laboratories, schools and offices should be calculated as farm area. Probably most extensive of all are the elaborate systems of roads, railroads, airports and docks which are as essential to a market-oriented farm as a plow.[12] Since these are used jointly, each farm benefits from the total space occupied by the infrastructure that supports it.

Spatial accounting of the type suggested here would involve enormous difficulties since it would require adding hectares of farm land to square metres of factory space or kilometres of road. Since I have never seen even a first approximation of such complete spatial accounting, I have no idea what the real productivity per unit of area of a modern farm might be (Stamp 1958). I suspect, however, that the advantages over a traditional farm are considerably less impressive than simple area-to-area comparisons suggest. In any event, the difficulties, perhaps even impossibilities of calculating spatial inputs, merely emphasise the inadequacy of unit are yield comparisons between traditional and modern farming systems.

I would hesitate to argue, as some do, that modern agriculture is no more productive than traditional farming. But lacking data on total labour and spatial inputs, there is no way to fully calculate how much gain is achieved. And once presumed higher productivity per unit of labour or land is brought into question, it can no longer serve as an automatic and indisputable argument of industrialising agriculture. Instead, additional goals such as employment, income distribution, and environmental protection must enter the calculus for determining which a mix of traditional and modern methods will produce the results desired by a particular society.

NOTES

1. Out of the plethora of terms used to differentiate farming systems, I have selected "traditional" and "non-industrial" and their opposites "modern" and "industrial" to contrast low-energy, labour-intensive with high energy, capital-intensive agriculture. For discussions of terminology see Kelly Harrison and Kenneth Shwedel, "Marketing Problems Associated with Small Farm Agriculture." Agricultural Development Council, Research and Training Network Seminar Report No. 5, November 1974; Mick Moore, "On Not Defining Peasants." (plus referenced literature), *Peasant Studies Newsletter*, Vol. 1, No. 4, October, 1972, pp. 156-58.
2. A. T. Mosher identifies four functional components of modern agriculture, of which farming is only one. The others are commercial agri-support activities, non-commercial agri-support activities, and the agrimilieu. *To Create a Modern Agriculture*. New York: Agricultural Development Council, Inc., 1971 pp. 5-7.

3. William Letwin points out that in many places agriculture is more capital-intensive and uses more power than manufacturing. "Four Falacies about Economic Development". *Daedalus*, Vol. 92, 1963, pp. 396-414.
4. Most methods of increasing modern farm productivity involve the use of purchased inputs.
5. "Indeed, energy in the form of human labour is, in the developed world, valued so highly that it is increasingly employed to perform a control function rather than as a source of direct energy input in most production processes." Vernon W. Ruttan, "Food Production and the Energy Crisis: A Comment." *Science*, Vol. 187, February 14, 1975, pp. 560-61.
6. For a review of the labour-quantity theory of value, see Joseph A. Schumpeter, *History of Economic Analysis*. New York: Oxford University Press, 1954.
7. For Example, Orville L. Freeman, *World Without Hunger*: New York: Frederick A. Praeger, 1968, p. 9. Data for this type of statement are drawn from standard statistical sources such as the Food and Agriculture Organisation (FAO) of the United Nations *Production Yearbooks* which minutely and very clearly define occupational types.
8. David Pimental, *et al.* "Food Production and the Energy Crisis." *Science*, Vol. 182, November 2, 1972, pp. 443-49.
9. Shifts of population from rural to urban areas often are linked to increasing farm industrialisation and productivity. They also should be related to transfers of support activities from rural to urban areas.
10. This is not to criticise the tendency to spread costs over the general society, but only the practice of making direct comparisons between farming systems that are unequally subsidised.
11. The situation has been described in caloric terms by Lamont Cole, ("Man's Ecosystem." *Bio-Science*, Vol. 16, 1966, pp. 243-48):

    > We boast of our efficiency in raising more food on less land with the labour of fewer people, but we are deceiving ourselves. If we were to deduct from the food calories that a farmer produces the calories consumed by his machinery, the calories used to build and transport that machinery, to mine raw materials, to process, transport and apply fertilisers and pesticides and to process and distribute the food, we should see that modern agriculture is largely a device for exchanging calories in fossil fuels for calories in food.

12. Students of shifting cultivation are criticised if they fail to include the extensive fallow areas required by these systems in calculating unit area productivity or carrying capacity. The argument, which is sound, is that continued production in field rotation systems depends upon the regenerative forces at work in fallow fields. (For a review, see John M. Street, "An Evaluation of the Concept of Carrying Capacity." *Professional Geographer*, Vol. 21, March 2, 1969, pp. 104-07). Advocates of industrialised farming should be equally rigorous. Who can deny the regenerative forces at work in a fertiliser factory or the contribution of the roads that bring these forces to the fields?

REFERENCES

Clifton, R. W., 1969 : "Subsistence Agriculture — Concept and Scope in Warton" (Ed), *Subsistence Agriculture and Economic Development*, pp. 19-20.

Cole, L., 1966 : "Man's Ecosystem" *Bioscience*, Vol. 16, pp. 243-48.

David, P., et al., 1972 : "Food Production and the Energy Crises", *Science*, Vol.182, pp. 443-47.

F.A.O., 1977 : *Production Yearbooks.*

Freeman, O. L., 1968 : *World Without Hunger*, New York.

Harrison, K. and Shwedel, K., 1974 : "Marketing Problems Associated with Small Farm Agriculture", Agricultural Development Council, *Research and Training Network Seminar Report No. 5.*

Howard T. O., 1971 : *Environment, Power and Society*, New York, pp. 115.

Letwin, W., 1963 : "Four Fallacies About Economic Development", *Daedauls*, Vol. 92, pp. 396-414.

Moore, M., 1972 : "On Not Defining Peasants", *Peasant Studies Newsletter*, No. 4, Vol. I, pp. 156-58.

Mosher, A. T., 1971 : *To Create Modern Agriculture*, pp. 5-7.

Schumpeter J. A., 1954 : *History of Economic Analysis*, New York, Oxford University Press.

Stamp, L. D., 1958 : "The Measurement of Land Resources", *The Geographical Review*, Vol. 48, No. 1, pp. 1-15.

Street, J. M., 1969 : "An Evaluation of the Concept of Carrying Capacity", *Professional Geographers*, Vol. 21, pp. 104-07.

2

IMPACT OF CASTE ON PRIMARY OCCUPATIONS IN INDIA
A Geographical Analysis

NOOR MOHAMMAD

Conceptual Framework

CASTE AND primary occupations are so closely linked to each other that they are quite often taken as synonyms. In the past every caste had an occupation of its own which used to be hereditary and to abondon it in search of others was considered to be sinful. But with the passage of time both have changed considerably. The purity of caste exists only in religious books. Inter-caste and inter-religion marriages and the cross breeds of their off-springs have resulted into innumerable castes. Similarly the occupational ramification was so rapid and diverse that several thousand of occupations are being adopted by inhabitants of the world. In fact during the twentieth century considerable technological changes have taken place leading to the improvement of overall economic condition of the people on the one hand and the socio-political upliftment of the masses on the other. Revolution has taken place in industrial and agricultural sector and the transport technology does not lag behind. The population has increased many-fold and the dietary habit of the people is changing fast. In dairy, poultry and piggeries several new breeds and birds are added which make them more profitable.

However, under such an overall caste-occupation dynamism, caste-occupation relationship at micro scale still exits. The degree of relationship depends, by and large, on level of technological development and modernisation. In backward and rural areas the primary occupations are still associated with certain castes. Since it is their hereditary occupation, they know all its techniques, methods, ins and out leading to higher production. We should, therefore, encourage the development of primary occupation according to the caste which have it on hereditary ground and the micro level planning should be done on this basis. Therefore, it is essential to study caste-occupation relationship for a rational and meaningful planning for the overall development of a region. For this purpose the state of Uttar Pradesh has been taken in the present study as it was not possible to collect the primary data for the entire country.

The Geographical Profile of the Region

The state of Uttar Pradesh, situated in the northern part of India between 23°52′-31° 28′ N. Lat. and 77°04′-84°38′ E. Long., covers an area of 294,362 square kilometres. It has 48 districts in 1931 which have now been split into 56 districts (Fig. 2.1). Lying in the Ganga plain, the entire state, with few exceptions, is a vast homogeneous plain mainly drained by the rivers Ganga, Yamuna, Gomati and Ghaghara. The state enjoys sub-humid climate with three well marked seasons: summer, rainy and winter. The soil is fertile alluvium deposited by the river channels. It has a predominantly agrarian structure. Agriculture engaging roughly 80 per cent of the total population directly or indirectly, is the main occupation of the people and it provides more than 50 per cent of state income. The state is not favourably endowed with mineral resources and its industrial base is also comparatively poor. It has a fairly good network of transport and communication, although there are a lot of variations in it. Uttar Pradesh is one of the densely populated states of the country and has a very complex population structure. It is more so in case of caste composition of population. More than hundred castes are found here but the major ones are Brahmin, Rajput, Jat, Shaikh, Kurmi, Scheduled caste, Kori etc. Since the state is at comparatively low web of the level of technological advancement, the association between caste and primary occupation and more glaring and strong but it has not been studied so far (Fig. 2.2).

Overview of Literature

The caste and its various parameters have been thoroughly studied at international, national and regional levels in spatio-temporal

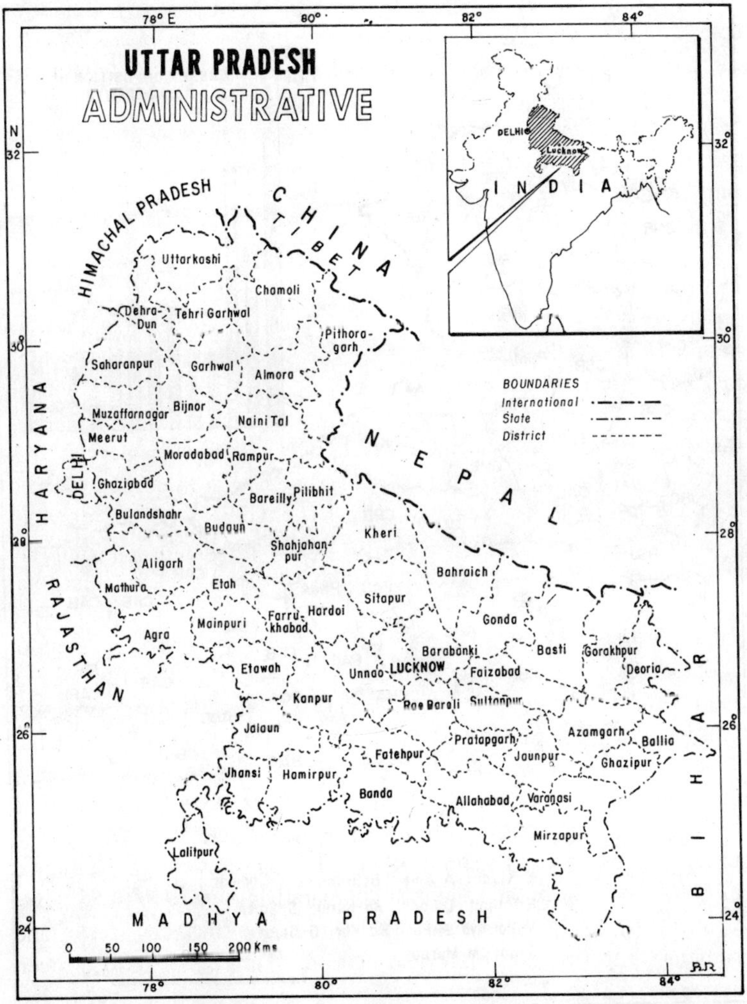

FIG. 2.1: Location of study area.

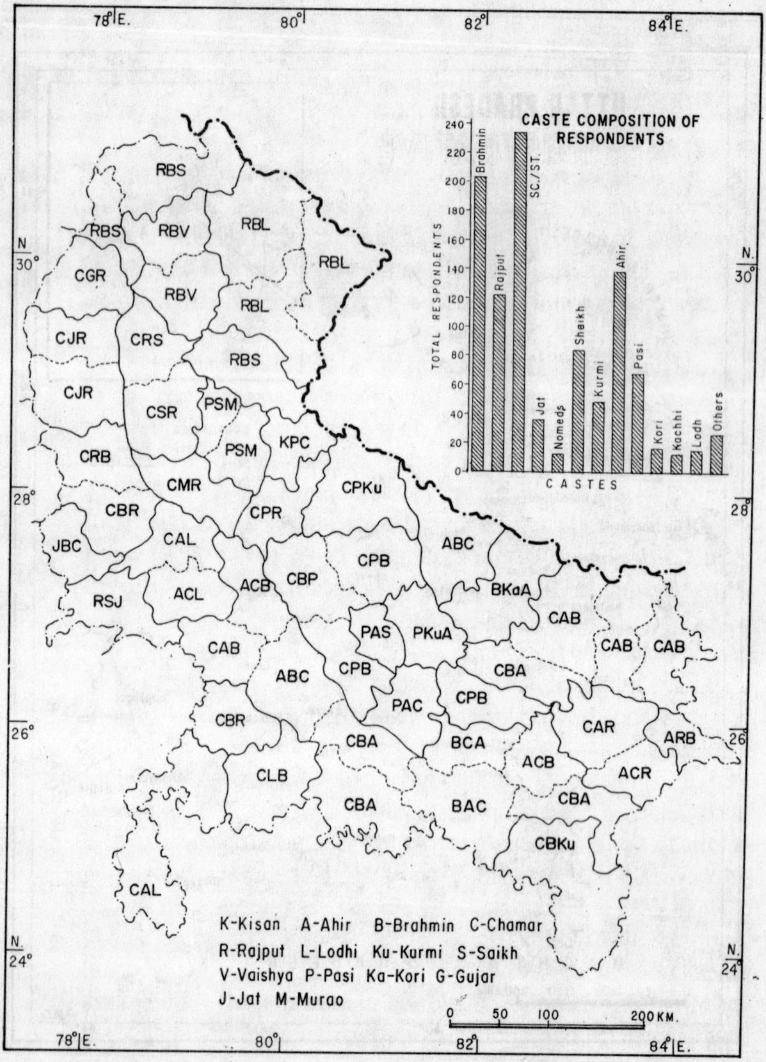

FIG. 2.2: Cast combination regions.

perspectives. The caste system, its nature, structure, has also been elaborately studied by sociologists, anthropologists and ethnographers. The caste, class and power have been one of the major themes of study of sociologists. Similarly various primary, secondary and tertiary occupations have also been a major theme of study for the geographers in particular, and economists and historians in general. The occupational structure of population has also been a point of study for geographers, demographers, economists, sociologists, and historians all over the world (Berry, et. al. 1978). Enumeration of occupation has been one of the major themes of the census organisations in almost all countries of the world.

Thus there are plenty of documents and literature on these two components viz. caste and occupation. But there is a dearth of studies and literature dealing with these two components together giving their inter and intra-relationships. Risley's (1915) study of caste composition and some of the major occupations at Indian census of 1901, is the pioneer work done on the national scale and under government aegis. But even his study is partially successful. Sociologists and anthropologists, although mainly concerned with the study of caste, class and power, have done some study in this respect from their own points of view (Omvetd, 1980). For example, in 1945 Gadgil in his first report of the Poona survey, has published his scheme of occupational categories based on their nature and status. His study of principal weaving castes of Maharashtra and the changing nature of their occupation, is one of the basic works of the caste-occupation relationship. Ghaurye (1961) has written a voluminous book on *Caste, Class, and Occupation* which sounds to be reasonably dealing with caste-occupation relationship. But it gives more specific treatment to caste and its correlates with a general chapter on caste and occupation. In this chapter too, a review of various studies dealing with occupation and occupational structure has been given with specific mention of few studies which deal with caste and occupation at micro or regional levels.

The geographers did study the primary occupations specially crop farming, dairy farming, vegetable farming etc. but they were busy in analysing the spatial distribution of agriculture with respect to physio-economic conditions (Singh, 1970, Mohammad, 1978, Symon, 1978), and the institutional bases of agriculture, especially caste, have not been studied in proper perspectives. The studies which are available, are sketchy and too general in this respect (Buchanan, 1969,

Symon, 1978, Mohammad, 1981 and Singh, 1985). Thus in geographical literature the study of caste and primary occupation is almost non-existant as far as the author knows. The author has given Field-Survey Projects on Impact of Caste on Primary Occupations to his M.A. final year students. The survey had been conducted in different villages located in the Union Territory of Delhi, and it gives different kind of association between caste and primary occupation which has prompted the author to conduct a thorough study to assess the relationship between caste and primary occupations.

TABLE 2.1: **Number of respondents from each caste**

S.No. Castes	No. of Respondents	%age to total
1. Brahmin	204	19.75
2. Rajput	123	11.91
3. Scheduled Castes	235	22.85
4. Jat	38	3.68
5. Nomads	14	1.36
6. Shaikh	85	8.04
7. Kurmi	49	4.75
8. Ahir	139	13.47
9. Pasi	69	6.69
10. Kachhi	14	1.36
11. Lodh	17	1.68
12. Kori	18	1.75
13. Other Castes	28	2.71
Total	1033	100.00

Data Source and Methodology

The study is essentially based on primary data collected from 1033 respondents of different castes chosen from 35 sample villages which have been taken from each caste combination regions worked out on Doi's method using 1931 data for caste, as it is the only secondary data available on caste (Table 2.1). These respondents have been interviewed according to a well structured questionnaire which besides containing a number of quarries, focuses on six primary occupations viz. crop farming, vegetable farming, dairy farming, poultry farming, fish farming and piggeries. Besides these, there are other occupations being followed by the people due to the variety of reasons, which have been broadly put into service, labour and other occupations (Fig. 2.3).

These Occupations have also been discussed along with the primary occupations. The data thus collected have been processed and analysed in order to examine and explain the association between the caste and the selected occupations. A detailed account of each selected caste and its occupation is given below.

Brahmins and their Primary Occupations

Brahmin-the levile class of Hindu caste system, who where once upon a time considered as priests, worshippers, composer or reciter of hymns, have varied occupations (Crooke, 1974). There are 204 total Brahmin respondents, out of which 162 or about 80 per cent have crop farming as their first occupation. The remaining 20 per cent have dairy farming, vegetable farming, service, labour and other occupations. The heavy dependence of Brahmins on crop farming is due to three reasons: firstly, their priestly occupation which was, of course, very remunerative in the past, has become almost obsolete; secondly, they have land which they have got, perhaps, from landlords and jagirdars as a result of performing religious, social and cultural rituals; and thirdly, the younger generation of Brahmins does not want to go by their hereditary occupation. Therefore, firstly they started crop farming by engaging people of lower castes as labourers, as ploughing by Brahmins was declared by themselves as against religion. But now with changed socio-economic conditions and non-availability of agricultural labourers, they have started ploughing the land themselves. There are 27 Brahmins or 18.12 per cent of the total 149 Brahmins who have crop farming as their second occupation, while only 4 Brahmins have crop farming as their third occupation. And as such, if Brahmins with crop farming as their first, second and third occupation, are put together, they account about 94 per cent of the total respondents, Except in poultry farming and piggeries, Brahmins have all the selected occupations. A comparative study of crop production and crop productivity of different castes indicates that Brahmin respondents are in a better position as compared to other castes. The production of wheat, rice and sugarcane among Brahmin respondents is much higher.

Vegetable farming is the other occupation engaging more than one-third of the total Brahmin respondents. About 4 per cent have vegetable farming as first occupation, while 17 per cent have it as their second occupation, although they account 24 per cent of Brahmins of

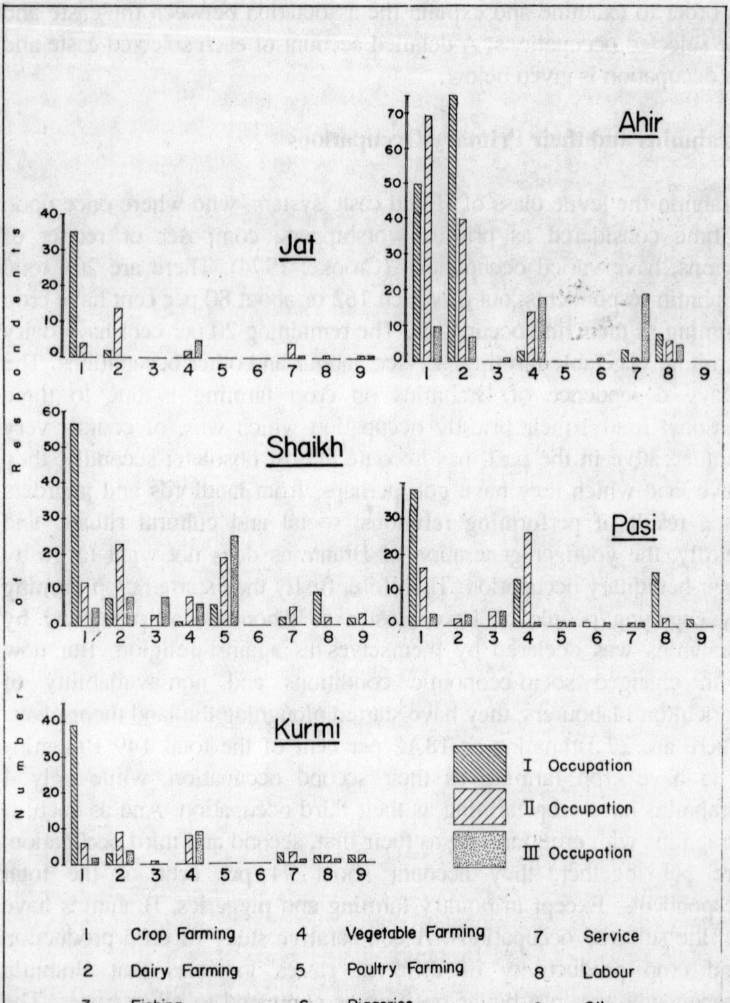

FIG 2.3: Caste-wise occupation.

this group. About 17 per cent of total Brahmin respondents or 34 per cent of Brahmins with vegetable farming as their occupation, have vegetable farming as third occupation (Table 2.2). The heavy concentration of Brahmins in vegetable farming is due to their dietary habits being governed by religious and socio-cultural laws, which do not permit them to be non-vegetarian. So vegetable constitute a major item in their diet. But the vegetable production is unsatisfactory among Brahmin respondents. The production of potato is, of course, higher among them.

TABLE 2.2: Brahmins and their primary occupations

Occupations	First No. of Respon.	Occupation % age	Second No. of Respon.	Occupation % age	Third No. of Respon.	Occupation % age
Crop farming	162	79.41	27	18.12	4	5.06
Dairy farming	5	2.45	39	28.17	22	27.84
Fishing	—	—	—	—	4	5.06
Vegetable farming	8	3.92	35	23.49	27	34.18
Poultry farming	—	—	—	—	—	—
Piggeries	—	—	—	—	—	—
Service	20	9.80	44	29.53	13	16.46
Labour	8	3.92	4	2.69	1	1.27
Others	1	0.50	—	—	8	10.13
Total	204	100.00	149	100.00	99	100.00

About 32 per cent Brahmin respondents have dairy farming as their first, second and third occupation. Five Brahmins or 2.45 per cent have dairy farming as their first occupation, while 39 Brahmin respondents have it as second occupation. They account more than 26 per cent Brahmins under this group. Twenty two Brahmin or 27.84 per cent Brahmins under third occupation, have dairy farming as their third occupation. Surprisingly 4 Brahmins have fishing as third occupation. In fact dairy farming, by and large, is not their basic occupation. They practice it on subsistence basis.

More than 37 per cent Brahmins are in service either having it as first, second or third occupation. It is a main occupation of about one-third of Brahmins, while secondary occupation of 21 per cent. About 7 per cent Brahmins are engaged in services as their third occupation. A scrutiny of questionnaire shows that Brahmins are, by and large, in white collar jobs. The poor Brahmins constituting about 6 per cent of the total, are labourers. Out of which for about 4 per cent, it is a main occupation, while for 4 respondents it is second occupation. Only one respondent considers it as his third occupation. Nine Brahmins accounting more than 4 per cent of the total, have miscellaneous occupations. This analysis shows that majority of Brahmins have crop farming followed by service and dairy farming.

Rajput and their Primary Occupations

Rajputs are considered to be warrior and ruling caste, and even to-day they constitute a major segment of Indian army. But however, due to changing socio-economic conditions, they are also changing their occupations. Crop farming is their major occupation engaging about 85 per cent of the Rajput respondents (Table 2.3). About 78 per cent Rajputs have crop farming as their first occupation. Only one Rajput has crop farming as his third occupation, while 20 Rajputs have it as second occupation. Due to several favourable socio-economic factors, the crop productivity among Rajput respondents is also high.

TABLE 2.3: Rajputs and their primary occupations

Occupations	First No. of Respon.	Occupation % age	Second No. of Respon.	Occupation % age	Third No. of Respon.	Occupation % age
Crop farming	96	78.05	20	21.74	1	2.27
Dairy farming	5	4.07	23	25.00	17	38.63
Fishing	—	—	2	2.17	3	6.83
Vegetable farming	2	1.63	23	25.00	12	27.27
Poultry farming	—	—	—	—	1	2.27
Piggeries	—	—	—	—	—	—
Service	16	13.00	21	22.83	10	22.73
Labour	4	3.25	3	3.26	—	—
Others	—	—	—	—	—	—
Total	123	100.00	92	100.00	44	100.00

Dairy farming is an other popular occupation among Rajputs engaging about 37 per cent of the total respondents. Only 5 Rajputs have it as first occupation, while 23 as second. Seventeen Rajputs have dairy farming as their third occupation, which is about one-third of the respondents under this category. Rajputs are not the professional dairy farmers. They keep it on subsistence basis. The overall production in dairy farming among Rajput respondents is low. Vegetable farming is almost equally popular among Rajputs accounting about 30 per cent of total respondents under it. Twenty three respondents have vegetable farming as their second occupation, while 12 and 2 respondents have it as second, third and first occupation respectively. Rajputs grow vegetables mostly for their·own consumption with few exception who sell it as well. Recently in the villages near the market centres, Rajputs have started growing potatoes on commercial scales. Except potatoes, the other vegetables grown by Rajputs give poor results, as they do not know the technique of doing it as well as they can not devote as much time as it requires. Although Rajputs are non-vegetarian, poultry farming and piggeries are almost absent among them except one respondent who has poultry farming as third occupation. Fishing is also not popular. Only 2 and 3 respondents have it as their second and third occupation respectively. This shows that fishing, poultry and piggeries are occupations adopted by low caste people, although its products are being consumed by high caste people. About 44 per cent Rajputs are in service and labour. The former engages 48 per cent, while the later only 6 per cent. The main occupation of 16 respondents is service, while 12 respondents have it as their second occupation. Like Brahmins, Rajputs are also in higher position with white collar jobs. The scrutiny of survey proforma indicates that they are in military, administration, and teaching. Only 4 Rajputs have labour as their first occupation, while 3 have it as second occupation. They are the poor and illiterate Rajputs.

Scheduled Castes and their Primary Occupations

Scheduled castes are popularly known as service and labour class since early times. In spite of all developments and modernisation, they still continue to be in the same occupation (Fig. 2.4). They, being the poor and down trodden, look for their livelihood all around and wherever occupation is available, they opt for it provided they are permitted to do so. Therefore, they are seen in all categories of the selected occupations. (Table 2.4). The highest percentage viz. 62 per cent is in crop farming followed by piggeries. Sixty nine respondents or 29 per cent of the total have crop farming as their first occupation, while 5.2 per cent respondents have it as second occupation. There and 24 respondents having it as third occupation. The latter two categories contribute 27 and 19 per cent of total respondents in their respective groups. Although a very percentage of scheduled castes have crop farming as their first, second and third occupation, their economic condition is unsatisfactory. The scrutiny of survey proforma indicates that they are marginal or very small size of farmers with limited financial resources. They have got land under the government's policy of distribution of land among landless. These lands were either *Banjar* lands of *Gram Samaj* or were acquired under Land Ceiling Act. In both the cases the quality of land is poor which is quite obvious because the land was left uncultivated and entered *Banjar* (infertile land) under *Gram Samaj*, and the farmers have given their worst quality of land in ceiling. But the scheduled castes who were deprived of land since ages happily accepted this land, and they are putting all their human and financial resources to get much out of it. But neither they have infrastructure nor the capital to built it. Therefore, the yield is very low. A time is likely to come very soon when these respondents will give up such lands and will go back to service and labour. But those respondents are doing better who have taken land from big peasants on lease basis or on share cropping basis because these lands are fertile and infrastructure is already built up there. Moreover, land owners advance credit for purchase of agricultural inputs.

TABLE 2.4: Scheduled castes and their primary occupations

Occupations	First No.of Respon.	Occupation % age	Second No. of Respon.	Occupation % age	Third No. of Respon.	Occupation % age
Crop farming	89	29.36	52	26.80	24	19.20
Dairy farming	9	3.83	10	5.15	5	4.00
Fishing	2	0.85	23	11.86	62	49.60
Vegetable farming	—	—	12	6.19	5	4.00
Poultry farming	—	—	4	2.06	8	6.40
Piggeries	72	30.64	56	28.87	9	7.20
Service	10	4.26	7	3.60	10	8.00
Labour	70	29.79	29	14.95	1	0.80
Others	3	1.27	1	0.52	1	0.80
Total	235	100.00	194	100.00	125	100.00

Piggeries their ancestral occupation, stands second engaging 58 per cent of scheduled caste respondents. Seventy two respondents or about 53 per cent of respondents have it as their first occupation, while 56 respondents or 41 per cent have it as their second occupation. Only

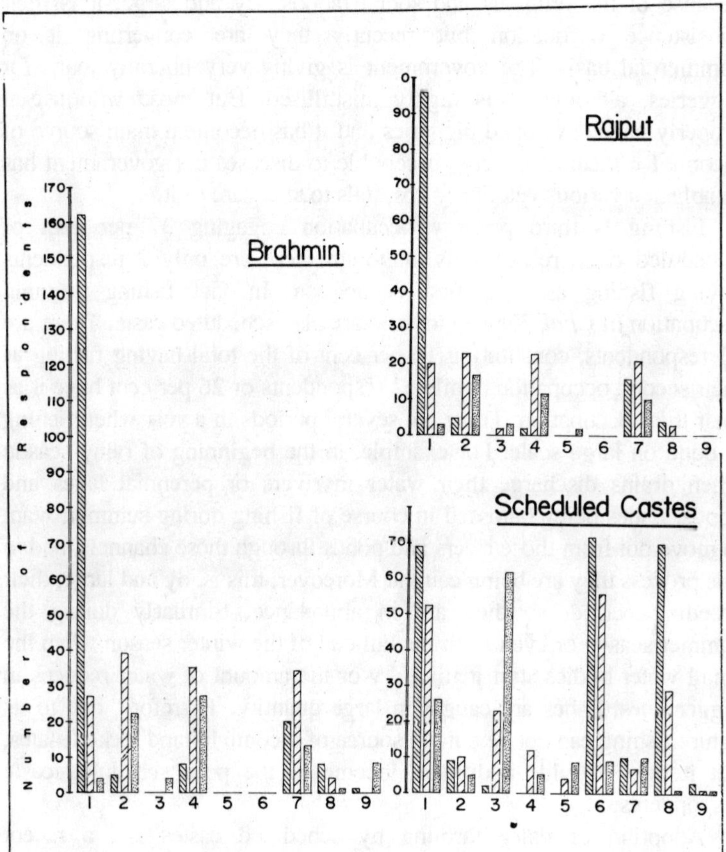

FIG. 2.4: Caste-wise occupation.

9 respondents have it as third occupation. Piggeries is their main occupation and no other caste except a few does this occupation because of the religious and social taboos. By and large, it is their subsistence occupation but recently they are converting it on commercial basis. The government is giving very liberally loans for piggeries, although it is highly misutilised. But those who use it properly have developed piggeries and it has become a main source of income for them. It is very susceptible to diseases but government has established various veterinary hospitals to take care of it.

Fishing is third primary occupation engaging 37 per cent of Scheduled caste respondents, although there are only 2 respondents having fishing as their first occupation. In fact fishing is main occupation of *Chai*, *Kewat* etc. who are also scheduled caste. There are 23 respondents, constituting 10 per cent of the total having fishing as their second occupation, while 62 respondents or 26 per cent have it as their third occupation. There are several periods in a year when fishing is done on large scale. For example, in the beginning of rainy season when drains discharge their water in rivers or perennial lakes and ponds, fishes being harassed in course of fishing during summer, want to move out from those rivers and ponds through these channels, and in this process they are being caught. Moreover, this is, by and large, their breeding period, so they are in abundance. Similarly during the summer season or even from the tail end of the winter season when the small water bodies start getting dry or the amount of water reduces in bigger ones, fishes are caught in large quantity. Therefore, due to its nature fishing can not be a main source of income in land locked states, but it does provide additional income to the poor scheduled caste respondents.

Adoption of dairy farming by scheduled castes is a recent phenomenon which is the gift of modernisation and socialisation. In the early times high and medium caste people could not even think of consuming the home made commodities of scheduled caste or any low caste. But now a days due to the development of the various service centres in the rural areas, tea stalls and sweet shops have come up like any thing which consume large quantities of milk. They purchase milk from every milk supplier irrespective of caste. Moreover, for high caste people, keeping and maintenance of dairy cattle has become a difficult task due to shortage of labour and shrinking of the pasture and grazing lands in which they use to leave their cattle free. Besides this, in the past the high caste people use to leave their cattle free in the village outskirts and no body could dare to tell anything to them against it. But now the conditions have changed and dairy farming requires large manpower to take care of it. Hence, scheduled caste, with large manpower, are venturing in this pursuit. Moreover, the government is providing loans with heavy subsidies to poor classes for dairy farming. However, their are 10 per cent respondents engaged in dairy farming. Out of which about 4 per cent have it as their first occupation, while almost equal percentage has is as second occupation. Five respondents have dairy farming as their third occupation. Majority of the scheduled castes dairy farmers sell the raw milk to these shops or sell it to dairy caste Ahirs, who supply it for home consumption to high and medium caste people. However, the dairy farming is giving a big economic boost to scheduled caste. Vegetable farming again has the same religious taboos because green vegetables grown and sold by scheduled caste will not be purchased by high caste people. But the conditions are changing and scheduled castes are also venturing in this occupation. There is none among scheduled caste with vegetable farming as his first occupation but there are 12 and 5 respondents having it as their second and third occupation respectively. Except few vegetables, they mostly grow it for their own consumption.

Large percentage of scheduled caste (42.55%) respondents are labourers having it as their first, second and third occupation. Labour is main occupation of about 30 per cent respondents, while about 12 per cent have it as second occupation. There is only one respondent with labour as his third occupation. Scheduled caste work as agricultural labourers in rural areas or as labourers in construction and manufacturing in town and cities. Recently with the taking up of several projects by government in rural areas like the construction of roads, embankments, canals, and several small scale industries, scheduled caste labourers have diverted their attention towards this. As a result, there is shortage of agricultural labourers which will adversely affect agriculture in coming decades, and eventually it would lead towards mechanisation. About 11 per cent of scheduled caste are in service also. While 10 each have it as their first and third occupation, only 7 have it as second occupation. Although the percentage of scheduled caste in service is reasonable, most of them are in low paid or inferior quality jobs. A very microscopic percentage is in white collar jobs. The survey schedule indicates that majority of scheduled castes are sweeper in urban centres or in clerical jobs.

Jats and their Primary Occupations

Jats are medium caste Hindus but occupy higher social status due to their better economic conditions and dominance in certain localities. They are sturdy and hard working. They are considered to be most efficient cultivators of India. They are slave to their farm (Russel and Lal, 1975). They have the feel of the soil and by having better agricultural infrastructure, they have high agricultural productivity. Their absorption is rigorous out of door work at all seasons, but it has had its effect on his character and physique (Crooke, 1974). Crop farming is their main and basic occupation. About 95 per cent Jat respondents have crop farming as their first occupation which speaks itself regarding favouritism of crop farming among Jats. A scrutiny of survey schedule indicates that they are, by and large, big and medium size farmers and do crop farming on commercial basis. There are only 4 respondents having crop farming as their second occupation (Table 2.5). Jats are very innovative and efficient farmers. Their technological knowledge and involvement is more, and they are economically better of also. Therefore, the crop production and crop productivity among Jat farmers is highest among all castes. They mainly produce wheat, jowar, bajra and sugarcane.

Dairy farming is the next important primary occupation of Jats. Ahir and Jat have close relationship, and in certain areas they are synonyms. However, only 2 respondents who did not have crop farming as their first occupation, have dairy farming as their first occupation, while 14 Jats contributing about 54 per cent of respondents under this category, have dairy farming as their second occupation. They do this occupation on commercial and subsistence basis. Majority of them sell raw milk in the villages or in the neighbouring market centres. Jat respondents are semi-professional in dairy farming. They keep better quality dairy cattle and overall dairy product among Jat respondents is high.

Vegetable farming is also popular among Jats but mostly it is done on subsistence basis. Generally they devote a small portion of their land to vegetable farming. A microscopic percentage of Jats sell vegetables either themselves through middleman. But recently potato cultivation with high yielding varieties has become more remunerative. It is more so as compared to wheat which requires higher inputs on one hand and government fixed price is very low on the other. Therefore, Jat cultivators have diverted their attention towards this. They have started devoting large areas under potato cultivation, and they sell it on commercial basis. But except potatoes, the production of other vegetables among Jats is low.

TABLE 2.5: Jats and their primary occupations

Occupations	First No. of Respon.	Occupation % age	Second No. of Respon.	Occupation % age	Third No. of Respon.	Occupation % age
Crop farming	36	94.74	4	15.38	—	—
Dairy farming	2	5.26	14	53.85	—	—
Fishing	—	—	—	—	—	—
Vegetable farming	—	—	2	7.69	5	71.42
Poultry farming	—	—	—	—	—	—
Piggeries	—	—	—	—	—	—
Service	—	—	4	15.38	1	14.29
Labour	—	—	1	3.85	—	—
Others	—	—	1	3.85	1	14.29
Total	38	100.00	26	100.00	7	100.00

None of the Jat respondents are in poultry farming and piggeries but about 19 per cent of them are engaged in service, labour and other occupations. The majority of their percentage is in service. The detailed scrutiny of the survey schedules indicates that they are in white collar jobs such as in military, teaching and administration. There is a single Jat respondent who has labour as his second occupation.

Shaikh and their Primary Occupations

In fact there is no caste system among muslims. In whatever from it is found among them, it is the gift of Indian soil. Among muslims there are two main sects — Sunni and Shia. Except these two, the other titles of muslims are on the basis of the community or leader, they are following, and within it also there are variation and complexities. For example, the Indian muslims who claim their linkages with 'quraish' — a community of Saudi Arabia to which Prophet Muhammad (Peace be on him) belonged, call them 'quraishi' or those who establish their linkage with Hazrat Farooq or Hazrat Siddique — the two calips, call them as Farooqi or Siddiqui respectively. But in India muslims are also divided on the basis of their occupation such as Quazi who use to be judge during the muslim rule. The caste of converted muslim is also according to their original Hindu caste. For example, Rajput converts call them Pathan or Khan but migrants of Afghanistan also call them Pathan. However, since the number of Muslim respondents was not large, all the muslim respondents have been put under one category and are called Shaikhs here. Therefore, whenever the term Shaikh has been used with reference to caste, it includes all the muslim respondents.

Since the muslims of all categories have been included in Shaikh, there is a wide range of occupation among them. Except in piggeries due to the religious bar, Shaikh have all the occupations selected for the present study (Table 2.6). Crop farming is quite popular among Shaikhs and 74 respondents or 87 per cent of total have it as their first, second or third occupation. About 67 per cent respondents have crop farming as their first occupation. It is their main source of livelihood. A reasonable percentage of Shaikh respondents are economically well of and have large size of holding. They produce wheat, rice, sugarcane and oilseed for commercial purposes. There are 12 respondents accounting 16 per cent of respondents in the category who have crop farming as their second occupation, while only 5 respondents have it as third occupation. Majority of the Shaikh respondents are very efficient crop farmer. With their moderate level of knowledge and involvement, they produce better crops of wheat, rice, and sugarcane. Crop farming is their hereditary occupation which also help them in increasing crop productivity.

TABLE 2.6: Shaikhs and their primary occupations

Occupations	First No. of Respon.	Occupation % age	Second No. of Respon.	Occupation % age	Third No. of Respon.	Occupation % age
Crop farming	57	67.07	12	16.00	5	9.62
Dairy farming	8	9.41	23	30.67	8	15.38
Fishing	—	—	3	4.00	8	15.38
Vegetable farming	1	1.18	8	10.67	6	11.54
Poultry farming	6	7.06	19	25.33	25	48.08
Piggeries	—	—	—	—	—	—
Service	2	2.35	5	6.67	—	—
Labour	9	10.59	2	2.66	—	—
Others	2	2.34	3	4.00	—	—
Total	85	100.00	75	100.00	52	100.00

Poultry farming is also very common among Shaikh engaging about 59 per cent of the respondents. Earlier it was considered to be women's job who use to do it as side occupation to support the family. By and large, it was common among low caste muslims. But now due to the changed dietary habit of large percentage of population, the poultry products are in great demand. Hence, in rural areas in general and in villages lying near towns in particular, poultry farming has gained popularity. Several young enterprising respondents have started there modern poultry farms, and poultry farming as domestic occupation has also considerably expanded. It has remained no longer a women's job but men are taking great interest in it and a large number of them are doing it.

There are only 6 respondents who are having dairy farming as their main occupation and they do it on commercial basis, while 19 respondents or more than one-fourth under this category have it as their second occupation. Still the larger number of Shaikh respondents accounting 29 per cent of the total respondents, are under this category. Dairy farming is favourite occupation of muslims who were basically nomads. It is a major source of income for Gujar, Gadaria, and Banjara who are, by and large, muslims. But among those muslims who are included in Shaikh, dairy farming is important primary occupation engaging about 48 per cent Shaikh respondents as their first, second and third occupation. There are only 8 respondent with dairy farming as their first occupation, while 23 respondents, accounting 31 per cent of the respective category have it as second and third occupation. Twenty seven per cent of total Shaikhs have dairy farming as their second occupation. In both the above categories, they do it on commercial and subsistence basis. There are 8 respondents who have it as their third occupation mainly on subsistence basis. Dairy farming is the main occupation of some of the respondents, and it is their hereditary occupation. But substantial number of Shaikh respondents do it on subsistence basis, and the quantity of dairy product is low.

Vegetable farming is not a favourite occupation of Shaikhs. They do it mainly on subsistence basis. There is a single respondent who has vegetable farming as his first occupation, while 8 and 6 respondents have it as their second and third occupation respectively. The only vegetable grown by the Shaikhs on commercial basis is potato. Shaikhs, being non-vegetarian, are also interested in fishing but it is not their main occupation. There is not a single Shaikh with fishing as his first occupation. There are three Shaikhs having fishing as second occupation, while 8 respondents accounting 15 per cent in this group, have fishing as their third occupation. They practice fishing on subsistence basis. The fishing is practiced by low caste muslims.

About 27 per cent Shaikhs are in service, labour and other occupations. Nearly 13 per cent are in labour and 8 per cent in service. They have these occupations as their first and second occupation. There are only two Shaikhs with service as their first occupation, while among 5 it is second occupation. Roughly 10 per cent Shaikhs have labour as their first occupation, while two have it as second occupation. A close look at the questionnaire shows that poor muslims are under these categories.

Kurmis and their Primary Occupations

Kurmi is a caste mainly found in central and eastern Uttar Pradesh. It is a very hard working efficient farming caste. The agricultural productivity among Kurmis is more. They, in fact, know how to conserve their agricultural resources. Crop farming is their main occupation, engaging about 96 per cent of the total Kurmis (Table 2.7). Out of this nearly 80 per cent have crop farming as their first occupation, while 6 and 2 respondents have it as their second and third occupation. They grow mostly wheat and rice and sell substantial portion of it. Dairy farming is an other occupation popular among Kurmis. There are only 3 respondents accounting nearly 6 per cent of the total respondents, who have dairy farming as their first occupation, but the larger number has it as second occupation. There are 9 respondents constituting nearly one-third of the respondents under this category. Four respondents have it as their third occupation. Kurmis are also known for vegetable farming especially onion and potato but incidentally it has not emerged as their first occupation. But about 35 per cent Kurmi respondents have vegetable farming as their second and third occupation. Eight respondents accounting nearly 29 per cent respondents, are in former, while 9 respondents accounting 53 per cent of the respondents, are in the latter group. They are, by and large, vegetarian caste and none of them practice poultry farming and piggeries. There is a single respondent having fishing as his third occupation.

Nearly 27 respondents are engaged in service, labour and other occupations. There are 3 respondents in service which is their main occupation, while one respondent has it as secondary occupation.

Kurmis are in better services. Labour is first and second occupation of 4 respondents — two falling in each category, while one respondent has it as his third occupation. Generally they are agricultural labourers with few exceptions.

TABLE 2.7: **Kurmis and their primary occupations**

Occupations	First No.of Respon.	Occupation % age	Second No. of Respon.	Occupation % age	Third No. of Respon.	Occupation % age
Crop farming	39	79.59	6	22.43	2	11.76
Dairy farming	3	6.12	9	32.14	4	23.53
Fishing	—	—	—	—	1	5.88
Vegetable farming	—	—	8	28.58	9	52.95
Poultry farming	—	—	—	—	—	—
Piggeries	—	—	—	—	—	—
Service	3	6.13	1	3.57	—	—
Labour	2	4.08	2	7.14	1	5.88
Others	2	4.08	2	7.14	—	—
Total	49	100.00	28	100.00	17	100.00

Ahirs and their Primary Occupations

Ahirs trace their descent from the God Krishna whose relations with milkmaids of Brindaban play an important part in Hindu mythology. Lord Krishna himself belonged to a nomadic caste who grazes cattle and makes butter (Risley, 1969). However, Ahirs are famous all over Uttar Pradesh for their dairy farming. It is their hereditary occupation. Among the selected Ahir respondents the similar trend has emerged. Out of 139 Ahir respondents 122 have dairy farming as their first, second and third occupation (Table 2.8). It accounts roughly 88 per cent of the total Ahir respondents. Nearly 54 per cent of total respondents have dairy farming as their first occupation, while 40 respondents, accounting more than 30 per cent respondents of the group, have it as their second occupation. There are 7 respondents having dairy as their third occupation. Ahirs are professional dairy farmers and it is their hereditary occupation. They are better knowledgeable and skilled persons in this respect. As a result the amount of dairy product among Ahir respondents is more as compared to the respondents of the other castes. However, by and large, they practice it on commercial basis. They sell raw milk as well as dairy products. In several villages they take out the butter and sell the skim-milk on barter system. The dairy farming, on the one hand, is expanding around the city centres, while on the other, it is declining in far away villages. Increasing demand of the dairy product is basic reason for the former, while deforestation and shrinking of grazing and pasture lands is the cause of the latter. But recently government has started giving highly subsidised loan for dairy farming which may give a boost to this occupation, although the loan is highly misutilised.

As a result of shrinking of grazing and pasture lands affecting dairy farming adversely, crop farming among Ahirs has gained popularity. Nearly 93 per cent Ahir respondents have crop farming as either first, second or third occupation. Fifty respondents have crop farming as their first occupation, while 69 respondents amounting to nearly 53 per cent of the respondents of the group, have it as their second occupation. An other 10 respondents, constituting nearly 17 per cent respondents of the group have it as their occupation. Ahirs grow variety of crops abut emphasis varies from region to region.

TABLE 2.8: Ahirs and their primary occupations

Occupations	First No. of Respon.	Occupation % age	Second No. of Respon.	Occupation % age	Third No. of Respon.	Occupation % age
Crop farming	50	35.97	69	52.68	10	16.67
Dairy farming	75	53.96	40	30.53	7	11.67
Fishing	—	—	—	—	1	1.67
Vegetable farming	3	2.16	14	10.69	18	30.00
Poultry farming	—	—	1	0.76	—	—
Piggeries	—	—	—	—	—	—
Service	3	2.16	1	0.76	19	31.67
Labour	8	5.75	6	4.58	5	8.32
Others	—	—	—	—	—	—
Total	139	100.00	131	100.00	60	100.00

Majority of Ahirs are vegetarian. Hence, they do not practice poultry, fishing and piggeries but there are two exceptions viz. one Ahir has fishing as his third occupation, while the other one has poultry farming as his second occupation. However, since they are basically vegetarian, more than one-fourth of Ahir respondents are engaged in vegetable farming as their either first, second or third occupation. Due to dominance of crop farming and dairy farming as first occupation, only 3 respondents have vegetable farming as their first occupation, but 14 respondents, accounting nearly 11 per cent respondents of the group, have vegetable farming as second occupation, while 18 respondents have it as their third occupation. They practice it on commercial as well as subsistence basis. They grow potatoes, onion and several kind of green vegetables.

About 30 per cent of Ahir respondents are in service and labour. The former covers nearly 16 per cent, while the latter 14 per cent. Service is main occupation of three respondents, while only one respondent has it as second occupation. There are 19 respondents, constituting nearly 32 per cent respondents of the group, with service as their third occupation. Some Ahirs are in military and administrative jobs, while majority is in clerical job. A few Ahirs are teachers as well. Nearly 6 per cent Ahir respondents have it as their second and third occupation respectively. They are either agricultural workers or freelance labourers doing whatever job they get at required time.

Pasis and their Primary Occupations

Pasi- a dravidian tribe, is a backward caste found mostly in central Uttar Pradesh. It is hard working, very efficient caste. There is, by and large, no religious and socio-cultural taboos in adoption of occupations. Hence, they emerged in all the selected occupations in this study. Crop farming is their major occupation engaging nearly 83 per cent of the total Pasi respondents (Table 2.9). Thirty eight respondents, accounting 55 per cent of total respondents have crop farming as their first occupation, while 16 respondents, amounting nearly 31 per cent of the respondents of this group have it as their second occupation. Only 3 respondents have it as third occupation. Pasis grow variety of crops and their crop productivity is high. Their entire family works in the field throughout the day and some times in the late evening and early morning.

Vegetable farming is also quite popular among Pasis. It engages nearly 58 per cent respondents. They are professional growers of vegetables and have age long experience in it. They know every detail of vegetable cultivation. Hence, the yield per unit of area and labour is high among Pasi respondents. They mostly grow green vegetables. However, it is main occupation of 13 respondents who account nearly 19 per cent of the total. Large number of Pasis have vegetable farming as their second occupation. There are 26 respondents, accounting 50 per cent respondents of this group, under this category. There is a single respondent with vegetable farming as his third occupation. Pasis grow vegetables mainly for selling purposes. Apart from potato and onion, they grow variety of seasonal green vegetables and sell it by themselves. They sell vegetables on cash as well as on barter system. Some times Pasis purchase vegetables from higher caste people and sell it.

TABLE 2.9: **Pasis and their primary occupations**

Occupations	First No.of Respon.	Occupation % age	Second No. of Respon.	Occupation % age	Third No. of Respon.	Occupation % age
Crop farming	38	55.07	16	30.77	3	20.01
Dairy farming	—	—	2	3.85	3	20.01
Fishing	—	—	4	7.69	4	26.67
Vegetable farming	13	18.84	26	50.00	1	6.67
Poultry farming	—	—	1	1.92	1	6.67
Piggeries	1	1.45	—	—	1	6.67
Service	—	—	1	1.92	2	13.30
Labour	15	21.74	2	3.85	—	—
Others	2	2.90	—	—	—	—
Total	69	100.00	52	100.00	15	100.00

Although fishing is not their main occupation, it is second and third occupation of nearly 12 per cent Pasi respondents. There are 4 respondents in each group but in the former it accounts nearly 8 per cent of this group, while in latter it constitutes about 27 per cent respondents in the group. They mainly do it on subsistence basis. Besides this, they are in poultry farming and piggeries as well. Piggeries is main occupation of one respondent, while the other one has it as his third occupation. Similarly in poultry farming there are only 2 respondents. One has it as his second occupation, while the other one as his third occupation.

Nearly 27 per cent Pasi respondents are in service, labour and other occupations but about one-fourth are labourers. About 22 per cent Pasis are labourers and it is their main occupation. They are either agricultural workers or freelancing labourers working in construction, earthwork, etc. However, their economic condition is poor. Only 2 respondents have labour as their second occupation, while there is none having it as third occupation. There are only 3 Pasis in service, having it as their second or third occupation. Out of it one is a school teacher, while the other two are peons.

Kachhis and their Primary Occupations

Kachhi is a caste similar to Kori or Pasi. It is a tribe of opium grower and market gardener (Crooke, 1974). It is a low caste, and the people are generally poor. Crop farming and vegetable farming is their main occupation. Nearly 57 per cent respondents have crop farming as their first occupation. There is little diversity in their occupation. There is none having crop farming as second or third occupation. Vegetable farming is second important occupation among Kachhis. They are professional vegetable growers and sellers. It is main occupation of one respondent, while 2 respondents have it as second occupation. They mostly do it for selling purposes. They sometimes purchase and sell vegetables. Dairy farming is not very popular among Kachhis but nevertheless 3 respondents amounting nearly one-fifth respondents, have it second or third occupation. Interestingly dairy farming is the only third occupation among Kachhis (Table 2.10). Nearly 29 per cent Kachhi respondents are labourers and it is their first occupation. They are either agricultural labourers or working in construction and earthwork. There is a single Kachhi in service and he is a peon in a local primary school.

TABLE 2.10: Kachhis and their primary occupations

Occupations	First No. of Respon.	Occupation % age	Second No. of Respon.	Occupation % age	Third No. of Respon.	Occupation % age
Crop farming	8	57.14	—	—	—	—
Dairy farming	—	—	1	33.33	2	100.00
Fishing	—	—	—	—	—	—
Vegetable farming	1	7.14	2	66.67	—	—
Poultry farming	—	—	—	—	—	—
Piggeries	—	—	—	—	—	—
Service	1	7.14	—	—	—	—
Labour	4	28.58	—	—	—	—
Others	—	—	—	—	—	—
Total	14	100.00	3	100.00	2	100.00

Lodh and their Primary Occupations

Lodh is a very low caste but they are very hefty and efficient. There is no religious bar in adopting any occupation except piggeries, but crop farming and fishing is their main occupation. Nearly 82 per cent Lodhs have crop farming as their first occupation which accounts about 59 per cent of total respondents (Table 2.11). They practice crop farming on subsistence basis and grow mostly foodgrains. There are 2 respondents with crop farming as their second occupation. A similar number has crop farming as their third occupation but they account one-third respondents under this group. Fishing is second major occupation, and it is the first occupation of nearly 12 per cent respondents, while about 30 per cent respondents have it as their second occupation. Three respondents, accounting 50 per cent respondents of the group, have fishing as their third occupation. It is their commercial cum subsistence occupation. They sell it in cash or through barter system. It provides major support to their living. There is a single respondent having poultry farming as his second occupation. About 65 per cent respondents are in labour and service. It is a poor caste. Hence, nearly 47 per cent respondents have labour either as their first or second occupation. Five respondents, constituting nearly 30 per cent of Lodh respondents, have labour as their first occupation, while 3 respondents constituting about 18 per cent of the group, have it as their second occupation. There are only 3 respondents in service. Out of it 2 have it as their second occupation, while one has it as third occupation. One is a school teacher, while 2 are peon in other organisation.

TABLE 2.11: Lodh and their primary occupations

Occupations	First No. of Respon.	Occupation % age	Second No. of Respon.	Occupation % age	Third No. of Respon.	Occupation % age
Crop farming	10	58.83	2	11.76	2	33.33
Dairy farming	—	—	2	11.76	—	—
Fishing	2	11.76	5	29.43	3	50.00
Vegetable farming	—	—	2	11.76	—	—
Poultry farming	—	—	2	11.76	—	—
Piggeries	—	—	—	—	—	—
Service	—	—	2	11.76	1	16.67
Labour	5	29.41	3	17.65	—	—
Others	—	—	—	—	—	—
Total	17	100.00	17	100.00	6	100.00

Koris and their Primary Occupations

Koris (Also spelled as 'Koeri, or Koari') are low caste Hindu and are similar to Kachhi of central Uttar Pradesh. They are mostly found in eastern Uttar Pradesh. They are known as vegetable grower. They are most efficient vegetable growers and it is their hereditary occupation. They devote a small portion of their land for vegetable farming, and it is called *Bari* or *Behra*. They remain busy in it throughout the year producing one vegetable or the other. However, 50 per cent of the total Kori respondents have vegetable farming as their first occupation, while 3 respondents have it as third occupation (Table 2.12). They are not only cultivate and sell vegetables but also purchase and resell in the villages and also in local weekly markets. It is their main occupation and source of livelihood. Crop farming is the second important occupation of Koris, being first occupation of 4 respondents or nearly 22 per cent of the total. There are 7 respondents, accounting 50 per cent respondents of this group, who have crop farming as their second occupation. There are 2 respondents with crop farming as their third occupation. Hence, crop farming altogether engages nearly 72 per cent of total respondents. Among Koris crop farming is subsistence type and they grow mainly food crops.

TABLE 2.12: Koris and their primary occupations

Occupations	First No. of Respon.	Occupation % age	Second No. of Respon.	Occupation % age	Third No. ot Respon.	Occupation % age
Crop farming	4	22.22	7	50.00	2	2.00
Dairy farming	1	5.56	2	14.29	6	60.00
Fishing	—	—	—	—	—	—
Vegetable farming	9	50.00	3	21.43	—	—
Poultry farming	—	—	—	—	—	—
Piggeries	—	—	—	—	—	—
Service	—	—	1	7.14	1	10.00
Labour	3	16.66	1	7.14	1	10.00
Others	1	5.56	—	—	—	—
Total	18	100.00	14	100.00	10	100.00

Dairy farming is also popular among Koris constituting 50 per cent of total respondents with either first, second or third occupation. There is a single respondent with dairy farming as his first occupation while there are 2 and 6 respondents having it as their second and third occupation respectively. They constitute nearly 14 and 60 per cent of respondents in their respective groups. They practice it on subsistence and commercial basis. Koris are purely vegetarian and they are locally know as *Bhagat.* Hence, fishing, poultry farming and piggeries are totally absent among them. And, perhaps, this is the reason that they are quite acceptable and popular vegetable grower and seller among all castes.

They are, by and large, poor. Hence, the large percentage work as labourer. Labour is main occupation of nearly 17 per cent respondents. There is one respondent with labour as second occupation and similar number has it as third occupation. Service is not popular among Koris. There are 2 respondents in service. One is a teacher, while the other is peon in the same school.

Nomadic Caste and their Primary Occupation

Generally *Gujar, Banjara* and *Gadaria* are nomads. They move with their flock of animals and all belongings from one place to other. But due to the technological change, modernisation and shrinkage of grazing and pasture lands, they have started sedentary life. However, due to their mobile nature, they have been put together and are called Nomadic Caste, although in census they are given separately. Their total number is also very small. Crop farming and dairy farming are their main occupations. The former is generally practiced by sedentary population, while the latter by Nomadic. Crop farming is the main occupation of 8 respondents who constitute nearly 57 per cent respondents, while a single respondent has it as second occupation. There are 8 respondents accounting nearly 43 per cent with dairy farming as first occupation, while there are 9 respondents constituting 60 per cent respondents of this group within it as their second occupation. A single respondent has it as third occupation. They are absent in vegetable farming, service and piggeries. There are 2 respondents in fishing and equal number is in poultry farming, having their second and third occupation (Table 2.13). Three respondents are labourers and it is their second occupation. They are engaged in various kind of labours.

Other Castes and their Primary Occupations

Besides above discussed castes, there are other castes such as

Vaishyas, Kumhar, Carpenter and Barber who are very small in number. Hence, they are grouped together under 'other castes' which together account about 2 per cent of total respondents. Crop farming and dairy farming are their main occupations. There is none in poultry farming and piggeries, while there are only 2 respondents in fishing — one each having first and third occupation (Table 2.14). Crop farming cngages 18 respondents accounting nearly 64 per cent of total respondents as first and second occupations. There are 15 respondents constituting nearly 54 per cent of total who have crop farming as their first occupation, while 3 respondents were having it as their second occupation.

TABLE 2.13: Nomadic castes and their primary occupations

Occupations	First No. of Respon.	Occupation % age	Second No. of Respon.	Occupation % age	Third No. of Respon.	Occupation % age
Crop farming	8	57.14	1	6.67	—	—
Dairy farming	6	42.86	9	60.00	1	33.33
Fishing	—	—	1	6.67	1	33.34
Vegetable farming	—	—	—	—	—	—
Poultry farming	—	—	1	6.66	1	33.33
Piggeries	—	—	—	—	—	—
Service	—	—	—	—	—	—
Labour	—	—	3	20.00	—	—
Others	—	—	—	—	—	—
Total	14	100.00	15	100.00	3	100.00

TABLE 2.14: Other castes and their primary occupations

Occupations	First No. of Respon.	Occupation % age	Second No. of Respon.	Occupation % age	Third No. of Respon.	Occupation % age
Crop farming	15	53.57	3	17.65	—	—
Dairy farming	4	14.29	7	41.18	6	50.00
Fishing	1	3.57	—	—	1	8.33
Vegetable farming	—	—	1	5.88	3	25.00
Poultry farming	—	—	—	—	—	—
Piggeries	—	—	—	—	—	—
Service	1	3.57	—	—	2	16.67
Labour	3	10.71	1	5.88	—	—
Others	4	14.29	5	29.41	—	—
Total	28	100.00	17	100.00	12	100.00

Dairy farming has a wider coverage. It is being adopted as first, second or third occupation by 17 respondents, while 4 respondents have it as first occupation. Seven respondents are practicing it as their second occupation accounting about 41 per cent respondents in the group. There are 6 respondents constituting half of the respondents of the group who have dairy farming as their third occupation. A large percentage of these castes are in service, labour and other occupations, which together engage nearly 54 per cent of total respondents. Out of this, only 3 are in service, 4 are labourers, and the remaining are in other occupations.

Conclusion and Suggestions

The following points clearly emerge out of the above discussion of caste and primary occupations:

(i) Crop farming is the main occupation in the state. The similar pattern emerges in the analysis of caste and primary occupations. Nearly 57.3, 21.2 and 5.1 per cent respondents have crop farming as first, second, and third occupations respectively. Crop farming is the main occupation of high caste respondents. They grow fine grains and more remunerative crops as they produce them for selling purposes. On the contrary, low caste respondents grow course grains and produce it on subsistence basis. Similarly there is a considerable difference in agricultural production also. The agricultural productivity, by and large, is higher among high caste respondents as compared to low caste respondents.

(ii) Dairy farming is next important occupation engaging 380 respondents with their either first, second, or third occupation. Out of this nearly 27.6 per cent have dairy farming as their first occupation, while 47.6 and 21.3 per cent have it as their second and third occupation respectively. Dairy farming, by and large, is main occupation of Ahirs, although the other castes also practice is but mainly on subsistence basis. Ahirs practice dairy farming on commercial basis. Nearly one-third of total respondents engaged in dairy farming as their first, second or third occupation are Ahirs. Among the Ahir respondents nearly 88 per cent are engaged in dairy farming as their first, second, or third occupation. Out of this nearly 54 per cent have it as their first occupation, while about 29 and 5 per cent have it as second and third occupation respectively. Even among Ahir respondents with dairy farming nearly 61 per cent have it as first occupation, and 33 and 6 per cent as second and third occupation respectively. Besides occupational hierarchy, there is variation in dairy products also. The production per unit of animal or per head is higher among Ahir respondents as compared to the respondents of the other caste who practice dairy farming. Ahir respondents know the art of keeping dairy cattle, whereas the respondents of the other castes lack it. He is also almost a semi-doctor of various diseases of dairy animals.

(iii) Vegetable farming ranks third in terms of number of respondents engaged. There are 259 respondents accounting 25.07 per cent of total respondents who have vegetable farming as first, second or third occupation. Nearly 14.29 per cent have vegetable farming as their first occupation, while 52.51 and 33.20 per cent have it as their second and third occupations respectively. Vegetable farming has got considerable changed especially around urban centres. People of different castes have started it. But still in rural areas, especially in remote parts, it is more popular among Kachhi, Kori, Kurmi and Pasi. Nearly 28 per cent of total respondent engaged in vegetable farming as their first, second and third occupation, belong to these castes. About 48 per cent respondents of these castes have vegetable farming as their first, second and third occupation. Out of this, nearly 32 per cent have dairy farming as their first occupation, whereas 54 per cent and 14 per cent have it as their second and third occupations respectively. The respondents of these castes know the minute detail of vegetable farming. And, therefore, the quality and production of vegetable in the fields of these respondents is better and more as compared to the respondents of the other castes.

(iv) By and large, poultry farming is practiced by Shaikhs and that too by low caste people. The other castes who practice poultry farming belong to lower castes. Nearly 59 per cent Shaikh respondents have poultry farming as their first, second or third occupation, and they constitute nearly 74 per cent of the respondents practicing poultry farming. However, among the poultry farming Shaikh respondents 12 per cent have it as first occupation, while 38 and 55 per cent practice it as their second and third occupation respectively.

(v) Piggeries is the main occupation of Scheduled Castes. It is rarely practiced by the respondents of the other castes. Nearly 58 per cent Scheduled Castes have piggeries as their first, second and third occupation. They constitute nearly 98 per cent respondents practicing this occupation. Among the piggeries practicing respondents, nearly 53 per cent have it as their first occupation, while 41 and 6 per cent have it as second and third occupation respectively. Fishing is common among low castes whether they belong to Hindu or muslim religion. Scheduled Castes, Lodh and low caste muslims have it as their first, second or third occupation.

(vi) The poor respondents of higher castes in general and the respondents of low castes in particular are labourers. They work as agricultural labourer, or freelancing labourers. The Scheduled castes dominate the scene. Nearly 43 per cent Scheduled Caste respondents are labourers having it as their primary, secondary or tertiary source of income. They constitute nearly 51 per cent of total labourers respondents. Among the Scheduled Caste labourer respondents, 70 per cent have it as main source of livelihood, while 29 and only one per cent have it as their secondary and tertiary source of income respectively.

(vii) Substantial percentage of respondents in service belong to higher castes especially Brahmin, Shaikh and Rajputs. A minor percentage of lower caste respondents are in service. A further analysis of respondents in service indicates that higher caste respondents are in superior and better jobs (white collar jobs), while lower caste respondents are in inferior and ministerial services.

3

HUMAN RESOURCE FORMATION IN A RURAL BLOCK OF MAGADH PLAIN

An Indicator of Rural Development

B. R. K. SINHA

A STUDY of Human Resource Formation (HRF) is currently realised extremely important for all round development of the region concerned. Out of the trinity of resources — physical, human and cultural — human resource is considered as an active agent of production which does not accumulate and exploite only nature's bountiful resources and increase the productive potential of a region but also increases and build social, economic, cultural, political strength and ultimately carry forward national development. In fact, human resource helps in the process of production of usable and consumable social values so that entire mass of the people can live a life under conditions of reasonable comfort after satisfying their basic needs. Undoubtedly levels of development in any form of an area depends largely on the levels of development of human resources, which is the result of HRF (a continuous and life long process of increasing knowledge, reasoning skills, energies, talents and other working capabilities of the people inhabiting the area).

The geographical area of the present study is Dhanarua — a rural Block of Patna District in Magadh Plain of Bihar State. It spreads over an area of 185.5. km^2 with the total population of 114971 (1981). Obviously the area is endowed with an abundance of human resources but the fundamental dilemma is that the majority of this remains undeveloped and under developed. Approximately two-third of the human resources remains unutilised. All these are directly or indirectly the result of human resources formation and impade social economic cultural and political development of the area.

Objectives

In the light of the importance of human resources and their associated problems the present work has been accomplished with the following objectives :

(i) to analyse and explain the present rate of human resource formation and to pin point the availability of those who are the off formation.

(ii) to suggest appropriate ways for fast development of human resource of the right type in the right quantity and quality at the right time for efficient and proper utilisation in the economic functions for rural development.

Methodology

Methodology consists of two parts — data base and student's age categorisation.

Data Base : The study is based exclusively on the primary informations collected from fourteen sample villages (eight cultivator Dominated villages and six Agricultural labourer Dominated villages) selected on the basis of stratified random sampling technique. From each village more than 20 per cent households were covered and informations regarding students' age, sex, classes in which they study and other were collected.

Students' Age Categorisation : Since age of an individual is very important for human resource formation, it involves some basic literature. Harbison and Myers (1964) considered three age groups for the students who are on human resource formation. First, 5–14 years which stands for those who are currently in the primary or elementary education, second, 15–19 years which comprises students of secondary education and third 20–24 years which includes the students of higher education. They considered such groups for international comparison however, these age groups vary from country to country because length of education varies according to needs and norms.

Ministry of education and social Welfare (1970-71) considered 6-11 years, 11-14 years, 14-17 years and 17-23 years as age groups for the pupils. This classification is also based on the length of schooling but this is very confusing. Age group of 6-11 years covers the pupils from six years of age while it has been found that some boys and girls who have completed five years of age are also studying in class/standard one. Third age group covers the students of four classes and length of schooling comes to only three years. Similarly last age group coves six years of the length of schooling while it should cover only five years.

Author has classified age groups into the following heads which are based on the length of schooling and classes covered by most of the students.

(i) 5-14, which includes the students from nursery and class/standard one to eight. This age group is based on the nine years of schooling instead of eight years because some boys or girls were reported in class one at the completion of the age of five years, while some at that age were reported to be in nursery class.

(ii) 14.1–18, which comprises four years of schooling and covers four classes from IX to XII.

(iii) 18.1–23, which stands for five years of schooling and covers degree and master courses. In professional/technical courses, only those were enrolled who had already completed either school or degree courses (formal education) but all of them fall in this age group.

In the study area human resource formation has been assessed on the basis of students' enrolment distribution in different classes under different age groups. Enrolment of students indicate the rate of HRF or Human Capital Formation (HCF) or Human Capital Accumulation (HCA) over specified period of time. Harbison and Myers (1964) also measured the rate of HRF on the basis of students' enrolment distribution under primary or elementary, secondary and higher levels of education. Todaro (1983) stated that the levels of accumulation of qualitative values depends primarily upon the methods of practice in schooling as schooling imparts specific knowledge and develops general reasoning skills and brings changes in belief and values and in attitudes towards work and society. Huq, (1976) also pointed out that schooling is a kind of human capital formation.

Analysis and Explanation

The following Table 3.1 clearly shows the difference in terms of the absolute figures and percentages of the enrolment of students in different age groups and classes between Cultivator Dominated (CD) and Agricultural Labourer Dominated (ALD) villages in the area.

Both the absolute figures and percentages of boys and girls (in the category of total students) in each age group are fairly higher in the CD villages than that in ALD villages. Secondly, both the number and percentage of students decrease from lower to higher age groups in both categories of villages. The same trend is found at the Block level as a whole and indicates the decreasing trend of enrolment of boys and girls towards the formation of better human capital which is indicated by higher education in higher age groups. This is in consequence of more drop-outs of boys and girls towards higher level of education for better quality formation.

Another observation is that the difference in number and percentage of boy and girl students increases with their increasing age groups. This trend indicates that the number of dropouts of boy and girl students is higher in each age group in ALD villages than that in CD villages. This is a consequence of their differential socio-economic background. The children in agricultural labourer's families become independent at their earlier age than that in cultivator's families. This indicates the sense of responsibility towards their family problems. It is also clear that the number of boys and girls is obviously higher in each class category and in each age group in CD villages than that in ALD villages.

In terms of percentage it is somewhat different as it is clear in the class category of I-V & VI-VIII for boys and girls respectively in the schooling period of 5-14 years. The percentage of boys is slightly higher in ALD than that in CD village. It may be because of their poor socio-economic conditions particularly in the case of boys which prevent them from joining higher classes and their percentage of drop-outs becomes higher at this stage. But the reason for becoming slightly higher percentage of girls in the class category of VI-VIII may be the growing awareness of the parents in sending their daughters to achieve slightly better education thinking that on the basis of this asset she would get better families to be married into. This is a good incentive in the process of HRF in the area.

In the schooling period of 14.1–18 years the percentage of boys is higher and of girls lower in CD villages as it is clear in the class category of IX–X, but the percentage of both boys and girls is higher in the class category of XI–XII in ALD villages than that in CD villages.

TABLE 3.1: Human capital formation (student enrolment, 1984)

Categories	Age Groups	Total Population		Total Students		Elementary Education I-WI-VIII		Secondary Education IX-X		Higher Sec. Education XI-XII		Degree Course		Master Course		Professional/ Technical			
		B	G	B	G	B	G	B	G	B	G	B	G	B	G	B	G	B	G
				201	115	122	81	73	29	6	5								
	5-14	305	254	(65.90)	(45.28)*	60.70+	70.43	36.32	25.22	2.99	4.35								
										39	7	29	3	4					
CD villages				72	10					54.17	20.00	20.28	30.00	5.56					
	14.1-18	119	64	(60.50)	(15.36)							5		24	1	4		3	
				36	1							13.89		66.67	100.00	11.11		8.33	
	18.1-23	106	82	(33.96)	(1.22)														
				126	69	83	44	40	22	3	3								
	5-14	196	173	(64.29)	(39.88)	(65.87)	63.77	31.75	31.88	2.38	4.35								
										15	4	17	3	1					
				33	7					45.45	57.14	51.52	42.86	3.03					
ALD villages	14.1-18	67	52	(49.25)	(13.46)									5		1		3	
				11								2							
												(18.18)		45.45		9.09		27.27	
	18.1-23	63	65	(17.46)															
				327	184	205	125	113	51	9	8								
	5-14	501	427	(65.27)	(43.09)	62.69	67.93	34.56	27.72	2.75	4.35								
										54	11	46	6	5					
				105	17														
At Block										51.43	64.71	43.81	35.29	4.67					
Level	14.1-18	186	116	(56.45)	(14.66)									29	1	5		6	
				47	1							7							
												14.89		61.70	100.00	10.64		12.77	
	18.1-23	169	147	(27.81)	(0.68)														

Source: *Author's Survey*, 1984.

* Bracketed Lower Figures are the Percentages of Boy & Girl Students to Total Population in Corresponding Age Groups.

\+ Unbracketed Lower Figures are the Percentage of Boys & Girls Enrolment to Total Boy or Girl Students in Corresponding Age Groups.

CD (Cultivator Dominated) & ALD (Agricultural Labour Dominated).

B (Boy) & G (Girl).

In the schooling period of 18.1–23 years the percentage of boys is slightly higher in the ALD villages as it is clear in the classes XI-XII and in the professional/technical classes. But their percentage in the degree and master courses are slightly lower. The higher percentage of boys in the professional/technical education indicates their growing tendency to be independent getting jobs at their earlier ages. The reason for this may be their inability to afford for higher formal education. Professional/technical skills help them in getting jobs easily even at their earlier ages.

It may also be seen at Block level that only 65.27 per cent boys and 43.09 per cent girls in the schooling period of elementary education (5–14 years) were on the HRF and 34.73 per cent boys and 56.91 per cent girls were off the HRF. Such a higher off percentages of boys and girls stand as an obstacle in the development of the area. In the schooling period of 14.1–18 years only 56.45 per cent boys and 14.66 per cent girls were on the HRF and 43.55 per cent boys and 85.3 per cent girls were off the HRF, the reason for which is the dropouts of the students in the higher classes. Similarly in the age group of 18.1–23 there were 27.81 per cent boys and 0.68 per cent girls on the HRF, whereas 72.19 per cent boys and 99.32 per cent girls were off the HRF. Such higher off percentages impede the development of human resources of higher level which are urgently and immediately needed for the development of the area.

HRF and Rural Development

Rural development is the process of increasing positive growth in agriculture, industry, trade and commerce, transport and communication, education, health, rural electrification, water supply, housing and in other basic needs of the people. Development of all these depends mainly on the rate of HRF or HRD. However, the development of human resources also depends on the economic, social, cultural, political, etc. development of the area. Even despite this, human resource is considered as an active agent and alter every aspect of life of the people by extracting, accumulating, transforming all types of resources and after all developing the total personality of the area and by making people with available essential resources for their basic needs to live a very peaceful, prosperous and satisfactory life. Thus all round development of the region is always linked with the balanced qualitative and quantitative development of human resources.

Harbison, Myers, (1964) stated that HRD or HRF is one of the necessary conditions for all kinds of growth — social, economic, cultural and political. Haq (1976) also stated that human capital formed by education has proved to be an active agent of economic growth.

Rural development is seen with the levels of literacy and education, levels of educational and health facilities including number of institutions, health status of the people, per capita income, participations rate of labour force in agricultural and non-agricultural economic functions, dependency ratio of women on men, dependency ratio of people on the agricultural and non-agricultural activities, age and sex structure, birth and death rates, child and senile labour utilisation, undeveloped and underdeveloped human resources and their unutilisation and underutilisation, movement of labour-force from agricultural to non-agricultural sectors and from one place to another, rigidity of caste, creed, party, religion etc. All these are the result of human resource formation.

Chandna and Sidhu (1980), stated that literacy is that qualitative attribute of population which is a fairly reliable index of the socio-economic development of an area and the lack of this keeps the population economically poor and mentally isolated. Literacy and education are also the indices of human resource development and human resource development itself is an indicator of rural development. It has been found that the percentage of literates and educated people is higher in the CD villages than that in ALD villages and the people of both categories of villages differ in terms of their socio-economic development. It has also been found that there has been an increase in the percentage of literates and educated people in the block as a whole which is a healthy sign for area's development. Similarly educational and health facilities including educational and health institutions have also been observed as indicators of rural development.

It has been found that the per capita income is higher in the group of qualitatively better developed human resources and lower in the group of undeveloped human resources. Therefore, human resource development can be considered as a reliable index of rural development.

It has also been found that there has been a slow shift in the participation rate of the working human resources from agricultural to non-agricultural economic function of the area. This has been observed as a consequence of HRD. Batra (1978), Aperyan (1979) have also stated that a shift or movement of manpower from agricultural to non-agricultural sectors is a general tendency of region's social, economic, cultural and political progress.

Work participation rate of women human resources has been found increasing from 1971 to 1981 and to 1984. This also has been because of an increase in the levels of their education. Evgeni Kapustin (1980) has also observed the increasing participation rate and of level of education of women in the agricultural and non-agricultural sectors of the economy, as a consequence of HRD.

The dependency ratio of females on males and the dependency ratio of the people as a whole on agriculture have been found decreasing which is a healthy sign of rural development and this also has been observed as a result of HRF.

Age is an index of individual's capability (the result of HRF) to work whereas sex ratio is an index of the socio-economic conditions prevailing in an area and is a significant element of regional analysis.

High birth and death rates indicates region's undeveloped and developing situations. High birth rate has been proved as the result of HRD. The National Sample Survey, 16th round (July 1960–June 1961, P. 4) explains that the number of children becomes smaller with higher educational standard of husband but the decline is particularly sharp with higher educational standard of wife. Birth and death rates have been found higher in the ALD villages than that in CD villages which may also be considered as the differential rate of HRF in these two categories of villages.

Very high proportion of males from the age group of 60 and above and child labour from the non-working age group (5–14 years) have been found working particularly in the agricultural activities. Females from both age groups are also working but their percentages are lower. Most of them are undeveloped human resources. This is not a healthy sign of area's development. Rigidity of caste, creed and religion has also been observed as impediment in the area's development. On the whole it is clear that HRF or HRD is an indicator or rural development.

Since HRF is considered as an indicator of rural development, there should be concrete planning for building the human resources of the right type in the right quantity and quality and at the right time which are essentially and urgently needed for region's economic, social, cultural and political development.

The formation or development of human resources of the above type can be possible through enrolment of all children falling under the age group of 5–14 years to achieve compulsory formal elementary education and through stopping drop-outs of students (particularly girl students) towards secondary, higher secondary, graduation and post-graduation including professional/technical education. This will bring all those to the process of HRF who are off the formation.

HRF can also be possible through Adult Education Programmes, on the Job Training Programmes and through the programmes conducted by several social, cultural, political, religions and other groups.

But in all cases the emphasis should be on vocational education as it develops practical ability opens avenues of job opportunities and contributes much to the basic needs of a developing economy. Thus functional education aimed at the masses and designed to mobilise and husband human resources, is a definite requirement for short as well as long range development of the area.

Games and sports including other activities are other important directions for developing human resources of the area.

Health is of course a wealth of a community which undoubtedly determines economic, social, cultural and political development of the area. Therefore, the provisions of medical facilities, health care, food and nutrition, potable water, recreational facilities and of the proper utilisation of working capacity of the people inhabiting the area in different economic functions are very essential for human resources development and rural development.

REFERENCES

Aperyan, V., 1979 : *Manpower Resources and Population Under Socialism,* Progress Publishers, Moscow.

Batra, V. P., 1978 : *The Economy of Human Resources,* B.R. Publishing Corporation, Delhi.

Chandna, R. C. and Sidhu, M. S., 1980 : *Introduction to Population Geography,* Kalyani Publishers, New Delhi.

Evgenie Kapustin, 1980 : Manpower in the U.S.S.R. : Development and Utilisation trends, in Paul Streetan and Harry Mair (Eds.) *Human Resources, Employment and Development,* Vol. 2, The Mc Millan Press Ltd., London.

Harbison, F. and Myers, C. A., 1964 : *Education, Manpower and Economic Growth : Strategies of Human Resources Development,* Mc Graw Hill Book Company, New York.

Huq, M. S., 1976 : *Education, Manpower and Development in South East Asia,* Steerling Publishers, Pvt. Ltd., New Delhi.

Ministry of Education and Social Welfare, Govt. of India, *Education in India,* (1970-71), New Delhi.

The National Sample Survey, Sixteenth Round (July 1960—June 1961) : No. 116, Tables with Notes on Family Planning, Published by the Manager of Publications, Civil Lines, Delhi.

Todaro, P. M. (Ed.), 1983 : The Struggle for Economic Development, *Readings in Problems and Policies* Longman, New York.

4

ROLE OF WOMEN IN AGRICULTURE

SHASHI KANTA VARMA

WOMEN ARE almost half of the adult population. They constitute one third of labour force but consume two thirds of the world's working hours and yet earn only one tenth of the income and own only one per cent of world property.

The prosperity and growth of a nation depend on the status and development of its women as they not only constitute nearly half of its population, but also influence the growth of the remaining half of the population. If a total range of women's actual economic activities are to be taken into account all over India, women work long hours in market and non-market economy, and more so in rural India. This draws our attention to the plight and drudgery of rural women.

This chapter (a) assesses the nature and extent of involvement of rural women and men of various socio-economic strata in farm, household and dairy activities in terms of operations, time spent, responsibility and accountability; (b) examines the comparative drudgery index of men and women for agricultural operations and relationship with socio-economic correlates and finally; (c) assess the women's exposure to modern crop technology and role of women in decision making in farming, home and other activities.

The multiple roles played and the productive inputs made by women in terms of work hours contributed or equivalent income generated in the family are neither attended nor recorded. According to the 1981 census, in India, there are 32 crore women out of the total population of 65.8 crore (48.6%). The primary sector consisting of farming, livestock, forestry, fishery, plantation and orchards etc., provides work to about 11 per cent of the total women population and about 87 per cent of the total women work force.

Women irrespective of land status of the family provide 14 to 18 hours of productive physical labour in different chores. The energy spent by them in performing these tasks are more than it is physically feasible for them to spend particularly in a below subsistence level of living.

Involvement of Women and Men in Farm, Household and Dairying Activities

Women are generally allocated menial jobs and often work for others as paid labourers. Research suggests that, across the centuries, the day-to-day life of rural women has been full of hardship, deprivation and the struggle to survive. For the women, life has meant that they are born to live as beasts of burden and to die without having lived one moment for themselves. Their life is meant to be lived for others, to be sacrificed at the altar of society's discrimination and allocation of a lower status to woman.

The operations viz., pesticide dusting, spade work during field irrigation, ploughing, sowing, manure and fertiliser application, uprooting of seedlings and marketing of grains are exclusively performed by men. Transplanting and storage of grains are found to be exclusive domain of women. The other farm operations like weeding, harvesting, carrying head load, threshing and winnowing are performed jointly but predominantly by women. Dixon (1982) also revealed that ploughing and irrigation were man's tasks, sowing may be restricted to men or women or shared, weeding and transplanting are women's tasks, harvesting is frequently shared and most of the post harvest operations (threshing, winnowing, drying, husking, cleaning, storing) are performed by women.

Chaudhary and Sharma (1961) revealed in their study that the rural women participate in agricultural operations like manuring, weeding, land preparation, hoeing, sowing, taking care of standing crops, applying manure and fertilisers, harvesting, threshing, storage, carrying the produce from farm to home. Similar findings have also reported by Naine, *et al.*, 1966; Singh, 1968; Devadas *et al.*, 1972; Bhatt, 1975; Dhar, 1978; Verma 1978; Devi and Reddy, 1984.

The average work load of men and women of low socio-economic stratum is higher than those of high and medium socio-economic strata respondents in farm operations. Women of low socio-economic stratum perform more work in crop cultivation than the women of high and medium socio-economic strata. This may be due to the fact that the respondents of low socio-economic stratum are mainly landless and wage earners. They are left with no other alternative for their livelihood, therefore, perform more work on other's field as agricultural labourers. Laxmi Devi (1987) reported that the low income category or rural women performed more work in farm activities while the high income category performed more work in household activities.

Working of family women on farm is valued lower and hence, better off families substitute family women labour force with hired labour force. It is only because of the economic compulsions that farmers of lower income allow their wives to work full time on the farm. This value goes against the dignity which needs to be given to farm work. A suitable educational programme to give dignity to the work they do, needs to be carried out with rural population.

Animal husbandry is predominantly a male affair in case of high socio-economic status because majority of them employ permanent male labour to look after the animals. Whereas, it is predominantly a female affair in case of farmers of medium and low socio-economic status. On an average, a woman devotes 3.5 hours per day for animal husbandry activities against only 1.6 hours per day devoted by men in these categories.

An analysis made by Puri (1975) shows that animal husbandry activities were done by women only. This was supported by Varma and Malik, 1984; Munjal, 1984; Kanjla, 1980; Kaur, 1987, Dhiman and Khirwar, 1986.

The household activities are predominantly performed by women irrespective of their socio-economic status and they devote more time in household activities during slack agricultural season. On an average, a woman devotes about 6 hours in peak season against 7 hours in slack season. Women devote more time in looking after the children and food preparation etc. during slack agricultural season but they tend to neglect these tasks during busy agricultural season. This is largely due to non-availability of agricultural labour. But the consequence of neglecting children and their food etc. is much more serious than the probable consequence on farm. This needs to be brought home through an educational programme and also, perhaps, mechanisation needs to be introduced to improve the efficiency on farm in peak season in particular so that curtailment of share of time for children is avoided.

For the male dominated agricultural operations responsibility lies with husbands only. The female dominated operations are the responsibilities of women in all socio-economic strata. The jointly performed operations have the joint responsibility of husband and wife in all socio-economic strata. Thus, the accountability is found to be the highest in case of jointly operated operations followed by female dominated operations. Male dominated operations has the lowest accountability.

Farm women involve themselves so much in farming but they do not have adequate knowledge of modern technology of crop cultivation. This is because there is no deliberate attempt made to give them education about modern technology. Even if some training programmes are organised, it is only the privilege of men to take part in such programmes. This practice needs to be put to an end and involvement of women in farm training programme have to be made to make agriculture moving.

Drudgery undertaken by men and women in cultivation of major crops (wheat and paddy)

Drudgery is generally conceived as physical and mental strain, fatigue, monotony and hardship experienced by human beings. While all of these results show decline in living and working conditions affecting men and women alike, the plight of the poor and the weak in this regard is alarming as they continued to be constrained by illiteracy, malnutrition and unemployment.

There is inter sex difference in the drudgery shouldered by men and women farm workers and also there is variation in the operations in terms of the amount of drudgery involved in them. In order to take account of this difference, the amount of drudgery taken up by every individual involved in one acre of wheat and paddy cultivation is presented in Table 4.1. The average drudgery data for men and women reveal that there is a significant difference in the total drudgery taken up by men and women in cultivation of wheat and paddy both. In both the cases, the drudgery load of women is more than that of men.

It is, however, worth noting that the results do not maintain consistency when average drudgery undertaken per man day of work in cultivation of wheat as well as paddy. Although the 't' value in case of both wheat and paddy cultivation are found to be significant, it is of interest to note that in case of wheat cultivation drudgery per man day taken up by women is more than that by men, whereas in case of paddy it is in the reverse order. Thus, it clearly reveals that on an average more drudgery prone tasks or operations are performed by women in case of wheat cultivation whereas men perform more drudgery ridden tasks in case of paddy cultivation. Yet the total drudgery suffered by women in both the cases are higher than men. This is due to the reason that women perform more hours of work than men in cultivation of both wheat and paddy.

TABLE 4.1: Drudgery taken up by men and women in cultivation of one acre of wheat and paddy

Crop		Average drudgery taken up	't' value	Average drudgery per man days	't' value
Wheat	Men	5.357	8.825**	0.598	6.763**
	Women	8.660		0.693	
Paddy	Men	7.948	4.396**	0.705	3.040**
	Women	11.199		0.662	

** Significant at 0.01 level of probability.

The results although do not appear to lend support to the generalisation that drudgery ridden agricultural tasks are assigned to women because of the male dominated social order, the fact that women shoulder greater total drudgery, deserves a particular attention. It is already mentioned that women shoulder almost entire responsibility of household activities and child rearing and also make a substantial contribution in dairying. In spite of all these, women work for significantly more hours than menfolk in cultivation of the two major crops — wheat and paddy shouldering greater total drudgery. This clearly points to the anormity of work load being carried out by women in the farming community. There is, therefore, a need to make the menfolk in particular, aware of the work load being carried out by women so that they may be more conscious of it and try to strike out a balance.

The assignment of different cultivation tasks to men and women are due to prevalence of two social values which play a significant role in it. They are : (a) women are physically less powerful than men and (b) they have lower ability to acquire skill. Men may be physically stronger than women in absolute term but what is ignored is the examination whether the amounts of power the particular task require are also possessed by women or not. This suggest a very serious implication to technological solutions if innovated to emancipate women from the drudgery they undertaken in farm operations/tasks. Tools and techniques may be devised to raise efficiency and reduce time and energy consumption in performance of the task assign to women. But then they may ungrudgingly hand over the task to men believing that they require higher level of skill.

In view of this Castillo (1977) made an startling revelation that even mechanisation in agriculture involves mainly those operations which are largely performed by men, while, women's work in both on and off the farm (including household tasks) has been predominantly manual. She further added that much of such technological innovations had absorbed male labour and disemployed female labour. This is supported with specific instance by Banerjee (1985) who said that with introduction of weedicides and pesticides, a large number of poor women lost their meagre source of income. According to him it is tautology that technological change and improvement in any sphere, trade and craft gives it status, increases economic return and make it the purview of men. Women meanwhile continue to cluster in the least skilled, most monotonous, most indispensable and low paid tasks in the production process.

Women's participation in agriculture is adversely affected by modern method of cultivation, which results in gradual displacement of women and shrinking of their activities. Similar conclusions were arrived at in the studies of Whyte (1978) in Java, Kabir and Chen (1978) in Bangladesh, Whyte and Whyte (1978) and Hanger and Maris (1973) in Kenya, Palmer (1977) and Nancy Bay (1970) in Africa and Dey (1985) in India.

Socio-economic Correlates of Drudgery

Drudgery reduces with increase in land holding due to reliance on hired labourers who share part of their drudgery. On the other hand, the casual labourers are the landless or marginal/small land holders whose drudgery increase because of their working on the farms of big land holders. The other reason is that in case of large land holding, different operations are carried out by own or hired tractors rather than bullocks which is exclusively done by men. So to that extent drudgery reduces as against farmers of lower holdings who carry out all such operations by bullock driven ploughs.

Farm women of lower income group use to undertake more drudgery load than the farm women of higher income group.

Decision-making Pattern in Farm Family

Men heavily predominate the decision-making process in farm family in all the six areas related to home management, child rearing, money management, socio-religious rituals, farm and animal management.

Domination of males in farm decisions has also been reported by the studies conducted by Puri (1972), Sharma and Singh (1970), Mulay (1974), Nagre (1976), Jhansi (1981), Yadav (1982) and Kaur (1987) at different points of time and cultural contexts.

This has made it amply clear that farm related activities are performed almost equally by the womenfolk as by the menfolk, but unfortunately their contribution in decisions related to farming is very meagre. Somehow the cultural value that 'Farm work is the domain of men and household of women' seems to prevail in the society without any regard paid to the contribution made by women in the family. This value seems to influence decisions about farm related activities taken in farmers' family, but this does not seem to influence the decisions regarding home management activities, where men are used to poke their nose and influence about half of the decisions. Even in case of other areas of decision-making domination of men is quite discernible. Yet, there is no sign of dissent from womenfolk against male domination. Male chauvinism obviously smacks from this. Therefore, there is a need to devise an educational programme for both men and women in the society to enhance respectability for women in farming and the society without disturbing the prevailing harmony.

Prestige Attached to Various Household Tasks

The prestige attached to the tasks performed by husband and wives clearly brings to light that in the eyes of women the tasks which they perform are not prestigious enough and the tasks which their husbands perform alongwith them or independently are significantly more prestigious. This speaks of the sense of inferiority, the farm womenfolk suffers from in our society. There is a need to pay attention to this so that the sense of inferiority goes from them. Without empowering them the dream of social growth and development can not be realised.

Conclusion

1. The various operations/tasks involved in cultivation of wheat and rice involve varying amounts/magnitudes of drudgery.
2. The farm operations/tasks are sex specific. Some tasks are performed exclusively by men and some other exclusively or predominantly by women. There are only a few tasks which are performed by men and women almost equitably.
3. The woman's involvement in agronomic operations/tasks in cultivation of crops is significantly higher than that of men in terms of man days of work done.
4. The total amount of drudgery in carrying out agronomic operations of rice and wheat cultivation, shouldered by women is significantly higher than that shouldered by men. They seem to be in direct proportion of the time spent in performing various tasks by men as well as women.
5. The amount of drudgery per man day of work shouldered by women is significantly higher than that shouldered by men in case of wheat cultivation, but it is in reverse order in case of rice cultivation.
6. The tasks assigned to men are those which are considered to demand greater physical power and skill for performance. The tasks assigned to women are those which are considered demanding less physical power, lower skill and more time, they are also monotonous in nature and low paid. The major force influencing inter-sex assignment of tasks is the social value held by men and women alike that women are *"abala"*, less powerful and have lower ability to acquire and use skills.
7. The farm machineries in use further demonstrate the play of this value. They are well being operated by men and not by women. This results in saving of time and energy of men giving them a greater leisure hours to rest, play cards and gossip. This also give some relief to women by relieving them of tasks like carrying head load etc. but it is quite marginal. This does not result in any unemployment to the womenfolk.

The conclusions reported above unmistakably brings to light that men are not callous to women. They consciously do not put women to hardships by assigning them the tasks of greater drudgery. In fact men and women both are unfortunate victims of some unrealistic, unreasonable and undesirable social values. An effort to innovate technologies to reduce drudgery involved in the tasks being performed by women is desirable, but it must accompany a well designed programme of education for both rural men and women to help them unlearn the unreasonable and undesirable norms and values and to help them inculcate in themselves the new values to achieve quality in life. Unless the two go together emancipation of rural women is unconceivable.

REFERENCES

Banerjee, J., 1985 : Implications of technology for women in the Rural Sector, in S. C. Jain, (ed.), *Women and Technology* Rawat Publications, Jaipur, pp. 7-16.

Bhatt, K. P., 1975 : A leading role in agriculture and animal husbandry. *Indian Farming,* 25 : 39.

Castillo, C. T., 1977 : *The Changing Role of Women in Rural Societies: A Summary of Trends and Issues, Seminar Report No. 12,* Agricultural Development Council, New York.

Census of India, 1981 : Provisional population totals, workers and non-workers. Series-I, Paper-3, India. pp. 31-59.

Choudhary, T. P. S. and Sharma, B. M., 1961 : Female labour of the farm family in agriculture, *Agricultural Situation in India,* 16 : 643.

Devadas, R. P., Thangamanik, S. Mithu, 1972 : Role of selected farm women in agricultural operations. *Indian Journal of Home Science,* 6 : 49.

Devi, L. A. and Reddy, S. V., 1984 : Role expectation and role performance of rural women in farm activities. *Indian Journal of Extension Education,* 20 : 27.

Dey, J., 1985 : *Women in Rice Farming System Network-Country profile on India,* IRRI, Philippines.

Dhar, A., 1978 : *Rural Women of India,* Kurukshetra. 26 : 4.

Dhiman, P. C. and Khirwar, A. S., 1986 : White revolution for welfare of rural women. Paper presented in the seminar on "Role of Development Programme on socio-economic status of women" held at Haryana Agricultural University, Hisar, from March 21-22.

Dixon, R. B., 1982 : "Women in Agriculture" : Counting the Labour Force in Developing Countries. *Population and Development Review,* 8 : 539.

Hanger, J. and Maris, J., 1973 : 'Women and household economy'; in Robert, C. and John, M (Eds.), *Mewar—An Irrigated Settlement in Kenya,* Munich, Weltforum.

Jhansi R. G., 1981 : *Decision-making Patterns among Rural Families and Type of Participation by Rural Women in Selected Areas of Decisions* (Unpublished M.Sc. Thesis), Andhra Pradesh Agricultural University, Hyderabad.

Kanjla, J., 1980 : A study of knowledge and extent of participation of women in dairy management practices (Unpublished M.Sc. Thesis), Punjab Agricultural University, Ludhiana.

Kaur, S., 1987 : *Women in Rural Development : A case study,* Mittal Publication, Delhi.

Laxmi Devi, A., 1987 : *Rural Women: Management in Farm and Home,* Northern Book Centre, New Delhi.

Mulay, S., 1974 : Role of farm women in Green Revolution, *Indian Journal of Adult Education*, 35 : 90.

Munjal, S., 1984 : An exploratory study of rural women's allocation, decision and activity pattern in Karnal district (Unpublished M.Sc. Thesis), Haryana Agricultural University, Hisar.

Nagre, M. L., 1976 : The role of family members in decision making as related to selected farm operations (Unpublished M.Sc Thesis), Marathwada Agricultural University.

Naine, M., Kherde, R. L. and Lokhande, M. R., 1966 : Performance of farm women towards food related tasks, *Indian Journal of Extension Education*, 4 : 35.

Nancy, H. J. and Bay, E. G., 1970 : *Women in Africa Studies in Social and Economic Changes*, Standard University Press, California.

Palmer, I., 1977 : Rural women and the basic needs approach to development. *International Labour Review*, 115 : 34.

Puri, S., 1972 : Work roles and decision-making pattern of farm wives and husbands. (Unpublished Ph.D. Thesis), Indian Agricultural Research Institute, New Delhi.

Puri, S., 1975 : *Role of Farm Women in Animal Husbandry Programme*, Kurukshetra 22 : 9.

Sharma, D. K. and Singh, T. R., 1970 : Participation of women in decision related to farm business, *Indian Journal of Extension*, Education. 6 : 43.

Singh, T. R., 1968 : Participation of rural women in agricultural operations in NES block, Jabalpur, (Unpublished M.Sc. Thesis), Jawaharlal Nehru Krishi Vishva Vidyalaya, Jabalpur.

Varma. O. S. and Malik, B. S., 1984 : Contribution of rural ladies in Farm Operations. *Indian Journal of Home Science*, 15 : 57.

Verma, J. C., 1978 : *Women in Rural Economy*, Kurukshetra, 26 : 7.

Whyte, R. O. and Whyte, P., 1978 : Rural Asian Women status and environment, Research notes and discussion paper No. 9, Institute of South Asian Studies.

Yadav, S., 1982 : Role of women in decision making related to farm business (Unpublished M.Sc. Thesis), Haryana Agricultural University, Hisar.

5

ABSORPTION OF HUMAN LABOUR IN TRIBAL AGRICULTURE

K. SATEESH REDDY AND K. MOHAN REDDY

THE RAPID decline of rural handicrafts and of the artisan class has added to the increase in the number of people dependent on agriculture and kept its proportion unchanged. As a result, the average real per capita income of the people living on agriculture has remained practically stagnant during the past 30 years or more. This should be a matter of great concern (Rath, 1989).

Agriculture is the kingpin of the tribal economy. The nature of tribal economy being predominantly dichotomous, i.e. agriculture and forestry forming its base, a significant percentage of the working force is engaged in these two sectors (Ramaiah, 1981). Nearly 80 per cent of the tribals live in rural areas and 93 per cent of them depend on agriculture and allied sectors. The agro-climatic conditions and the resource potential, however, vary considerably from one area to another.

This paper examines the absorption of human labour, male and female, at different levels of agricultural operations in the cultivation of three major crops, namely paddy, jowar and gingelly in Warangal district of Andhra Pradesh.

Profile of the District

The total cropped area of the district forms 38.4 per cent of the total geographical area. Of the cropped area, 31.3 per cent is irrigated. The principal crops grown in the district are paddy, jowar, maize, pulses and groundnut. The important irrigation sources in the district are Ramappa, Pakhal and Lakhnavaram and Salivagu project under which considerable area is irrigated. The other sources of irrigation are rain, tanks, wells and hill-streams which require good showers in the season. According to population (1981 Census) the district occupies twelfth place in the State. The total number of main workers accounts for 42.87 per cent of the total population and out of the main workers, cultivators and agricultural labourers together claim 72.46 per cent.

Methodology

The major objective of the study is to analyse the different levels of labour absorption, tribal and non-tribal, at different levels of agricultural operations such as ploughing, sowing, weeding, harvesting, threshing, etc., for different crops in Warangal district of Andhra Pradesh. The other aim is to examine the male and female participation in different operations with a view to examining the role of women in the agricultural sector.

For the purposes of the study, primary data were collected through the structured interview schedule. The data are also obtained by the participant observation method during both the seasons, i.e. Khariff and Rabi, at different operations involved in agriculture. This is done in order to ascertain the reliability and accuracy of the data. Random sampling technique has been adopted to select three Mandals, nine villages (three villages from each Mandal) and 153 households. The three Mandals are: Eturnagaram, Tadvai and Kothagudem. On the whole nine villages are selected for the study. From each village 10 tribal households and seven non-tribal households are chosen. Altogether 90 tribal households (30 from each Mandal) and 63 non-tribal households (21 from each Mandal) are selected.

Primary data is obtained pertaining to labour absorption from the sample households who are cultivating different crops such as paddy, jowar, gingelly, groundnut, maize, cotton, green-gram, red-gram, chillies, etc. However the present paper concentrates on three major crops, namely paddy, jowar and gingelly.

Analysis

The total labour force can be divided into owned labour and hired labour. An examination of the absorption of labour in paddy crop in Warangal district reveals that out of 153 sample households selected 142 households cultivate paddy. In this more than three-fifths (61.27

per cent, i.e. 87 out of 142 households) of the households belong to the Scheduled Tribes. Out of the total labour force absorbed for the cultivation of paddy crop, Scheduled Tribe people represent less than 50 per cent whereas majority of the labour absorption belong to non-tribal groups. Female labour force is more used in paddy cultivation. Table 5.1 shows the different levels of labour absorption in paddy crop at different levels of agricultural operations such as ploughing, sowing, weeding, harvesting, threshing and others. A large number of labour is used for sowing, weeding and harvesting when compared to ploughing and threshing.

An interesting feature is that female labour force dominates in the cultivation of paddy. Out of 1,932 labour force used for sowing, female labour force account for 1,652 (i.e. 85 per cent of the total labour force). As is well known, male domination is clearly visible in regard to ploughing. Male labour is more absorbed for threshing when compared to female labour. The ratio of female labour force to male labour force is higher in the case of non-tribals when compared to that of tribals. It is revealed that 87 tribal households who cultivated paddy during the period of study has used less number of labour force (2,904) whereas 55 non-tribal households have absorbed a higher number of labour force (3,179). Therefore, in the cultivation of paddy it may be concluded that labour absorption is more in regard to non-tribals when compared to that of tribal households. If one compares the absorption of tribal labour to that of non-tribal labour at different levels of operations, a wide difference can be observed in the operations of weeding and harvesting. Less number of tribal labour force was absorbed in these two operations whereas more tribal labour force was utilised for threshing and other purposes. Marginal differences can be seen in regard to ploughing and sowing.

An analysis of absorption of human labour in the cultivation of jowar and gingelly crops indicates three major features:

1. Absorption of hired labour is minimum.
2. Absorption of tribal labour seems to be more than the non-tribal labour. But the ratio of labour force absorption between the tribals and non-tribals is less when compared to the ratio of tribal households to non-tribal households who cultivated jowar and gingelly.
3. In comparison to male labour force, absorption of female labour force in the cultivation of jowar and gingelly is lower.

TABLE 5.1: Labour absorption in paddy crop

Operation and Caste Group	No. of days	Owned Labour			Hired Labour			Total Labour Force		
		Male	Female	Total	Male	Female	Total	Male	Female	Total
1. Ploughing		257	3	260	35	—	35	292	3	295
Tribals	1051	132	2	134	7	—	7	139	2	141
Non-Tribals	919	125	1	126	28	—	28	153	1	154
2. Sowing		258	236	494	22	1416	1438	280	1652	1932
Tribals	265	144	133	277	9	670	679	153	803	956
Non-Tribals	159	114	103	217	13	746	759	127	849	976
3. Weeding		262	244	506	15	1072	1087	277	1316	1593
Tribals	333	143	142	285	5	447	452	148	589	737
Non-Tribals	177	119	102	221	10	625	635	129	727	856
4. Harvesting		258	234	492	11	899	910	269	1133	1402
Tribals	302	140	131	271	—	301	301	140	432	572
Non-Tribals	171	118	103	221	11	598	609	129	701	830
5. Threshing		251	227	486	145	23	168	404	250	654
Tribals	251	131	125	264	80	7	87	219	132	351
Non-Tribals	233	120	102	222	65	16	81	185	118	303
6. Others		102	90	192	17	9	26	119	99	218
Tribals	212	58	55	113	8	6	14	66	61	127
Non-Tribals	123	44	35	79	9	3	12	53	38	91
7. Total		1411	1034	2445	242	3396	3638	1653	4430	6083
Tribals	2414	776	588	1364	109	1431	1540	885	2019	2904
Non-Tribals	1782	635	446	1081	133	1965	2098	768	2411	3179

Source: Field Study.

TABLE 5.2: Labour absorption in jowar crop

Operation and Caste Group	No. of days	Owned Labour			Hired Labour			Total Labour Force		
		Male	Female	Total	Male	Female	Total	Male	Female	Total
1. Ploughing		81	6	87	—	—	—	81	6	87
Tribals	161	47	2	49	—	—	—	47	2	49
Non-Tribals	109	34	4	38	—	—	—	34	4	38
2. Sowing		78	78	156	—	1	1	78	78	156
Tribals	52	44	41	85	—	—	—	44	41	85
Non-Tribals	34	34	37	71	—	1	1	34	38	72
3. Weeding		—	—	—	—	—	—	—	—	—
Tribals	—	—	—	—	—	—	—	—	—	—
Non-Tribals	—	—	—	—	—	—	—	—	—	—
4. Harvesting		88	87	175	—	7	7	88	94	182
Tribals	82	54	52	106	—	—	—	54	52	106
Non-Tribals	51	34	35	69	—	7	7	34	42	76
5. Threshing		80	82	162	2	—	2	82	82	164
Tribals	65	47	47	94	—	—	—	47	47	94
Non-Tribals	44	33	35	68	2	—	2	35	35	70
6. Others		21	17	38	—	—	—	21	17	38
Tribals	26	10	9	19	—	—	—	10	9	19
Non-Tribals	24	11	8	19	—	—	—	11	8	19
7. Total		346	263	609	2	8	10	348	271	619
Tribals	386	202	151	353	—	—	—	202	151	353
Non-Tribals	262	144	112	256	2	8	10	146	120	266

Source: Field Study.

Let us first explain the human labour absorption in the jowar crop. Labour is not required for the process of weeding. On the whole, the absorption of hired labour is marginal (1.6% only). In the case of tribal labour force, no labour was hired for the cultivation of jowar crop whereas only 10 non-tribal labour were hired. Except for ploughing, the absorption of female labour force is either equal to or more than the male labour force in the threshing, sowing and harvesting. In all the operations of jowar cultivation, more number of tribal labour force was absorbed. Table 5.2 clearly explains this situation. However, it is to be noticed that more than three-fifths (62.07%, i.e. 36 out of 58) of the sample cultivating households are tribal, while the rest, i.e. 22 out of 58 households, are non-tribal. Therefore, if one takes into account the size of tribal households cultivating the jowar crop, the ratio of tribal labour to non-tribal labour (57%) becomes less than the ratio of tribal households to non-tribal households (62.07%) cultivating the crop.

An examination of labour absorption with regard to the cultivation of gingelly crop reveals that it is cultivated by 27 tribal households and 20 non-tribal households. The sample cultivators of gingelly could not absorb many hired labour in the process of cultivation of gingelly. The percentage of hired labour to the total labour force is less than one (5 out of 627). Hired labour to the extent of one each were hired for ploughing, sowing and threshing, and two labour for harvesting (Table 5.3). On the whole, three tribal female labour and two non-tribal male labour were hired. Not a single female labour, either tribal or non-tribal, was absorbed for ploughing. The absorption of female labour for other operations was less than that of male labour force. If we analyse caste-wise absorption, more of female tribal labour was absorbed for weeding and other purposes when compared to male tribal labour. An equal number of female and male tribal labour force was absorbed in the operation of harvesting whereas less number of female tribal labour was absorbed in the operations of sowing and threshing when compared to male tribal labour. However, in the case of non-tribal labour force majority of them are male of all operation levels.

A highest number of labour was absorbed during the harvesting and threshing periods (146 and 141, respectively) and a minimum number during the ploughing period (67 only).

TABLE 5.3: Labour absorption in Gingelly crop

Operation and Caste Group	No. of days	Owned Labour			Hired Labour			Total Labour Force		
		Male	Female	Total	Male	Female	Total	Male	Female	Total
1. Ploughing		66	—	66	1	—	1	67	—	67
Tribals	93	35	—	35	—	—	—	35	—	35
Non-Tribals	83	31	—	31	1	—	1	32	—	32
2. Sowing		61	52	113	—	1	1	61	53	114
Tribals	34	33	28	61	—	1	1	33	29	62
Non-Tribals	25	28	24	52	—	—	—	28	24	52
3. Weeding		42	40	82	—	—	—	42	40	82
Tribals	25	22	23	45	—	—	—	22	23	45
Non-Tribals	19	20	17	37	—	—	—	20	17	37
4. Harvesting		74	70	144	—	2	2	74	72	146
Tribals	43	40	38	78	—	2	2	40	40	80
Non-Tribals	37	34	32	66	—	—	—	34	32	66
5. Threshing		72	68	140	1	—	1	73	68	141
Tribals	44	39	37	76	—	—	—	39	37	76
Non-Tribals	35	33	31	64	1	—	1	34	31	65
6. Others		39	38	77	—	—	—	39	38	77
Tribals	44	22	23	45	—	—	—	22	23	45
Non-Tribals	25	17	15	32	—	—	—	17	15	32
7. Total		354	263	622	2	3	5	356	271	627
Tribals	288	191	149	340	—	3	3	191	152	343
Non-Tribals	230	163	119	282	2	—	2	165	119	284

Source: Field Study.

Conclusions

It is found that the absorption of hired labour is more than that of own labour for the cultivation of paddy, whereas the absorption of hired labour is marginal for the other two crops, viz. jowar and gingelly. Another interesting feature is that the absorption of female labour is high in the case of paddy cultivation when compared to jowar and gingelly. The absorption of tribal labour in absolute terms is more for growing jowar and gingelly crops, whereas tribal labour force is less absorbed for the cultivation of paddy when compared to non-tribal labour force. But if we consider the ratio of labour absorption in the two categories to the ratio of cultivators in the same categories, the absorption of tribal labour force is less.

REFERENCES

Nilakantha Rath, 1989 : "Agricultural Growth and Investment in India", *Journal of Indian School of Political Economy*, Vol. I, No. 1, Jan.-June, p.64.

Ramaiah, P., 1981 : *Tribal Economy of India*, Life and Light Publishers, New Delhi, p. 32.

Hand Book of Statistics, Warangal District, 1986-87, compiled and published by the Chief Planning Officer, Warangal.

6

FUNCTIONAL CLASSIFICATION OF COMMUNES IN POLAND

WLADYSLAWA STOLA

THE STUDIES on the spatial structure and functional classification of rural areas in Poland conducted so far in the Institute of Geography and Spatial Organisation, Polish Academy of Sciences have been aimed at identification of the functional structure of these areas and elaboration of the principles and methods of their classification. The studies were conducted on the scale of voivodship rural areas and on the scale of communes in the selected voivodships (nearly 20 per cent of Poland's territory). The obtained results (Stola, 1987) have shown a considerable complexity and spatial differentiation of the functional structure and resulting from it a need to carry out detailed studies (on the scale of communes) on all rural areas in Poland.

Functional classification of communes is possible due to the fact that the studies conducted so far have solved the methodological problems pertaining to : (1) the subject matter of the studies; (2) a set of diagnostic attributes representing the particular functions and measures of these attributes; (3) the methods of normalisation of measures and groupings of the examined units.

From the point of view of the applicability of the results in the conducted studies and of their future utility, it has been assumed that rural areas are the territories of communes together with the small towns (capitals of communes). It means that a basic investigated unit is a rural and urban-rural commune. The basic (external) functions of rural areas in Poland, both in the meaning of the frequency of their occurrence and orientation towards the satisfactions of external needs (out side the investigated unit) are the following functions : the bioproductive functions — commercial agriculture and forestry; the technoproductive functions — industry and building; the services — tourism and rest (recreation) and a dormitory (residential) function.

While researching the problems we have selected a set of attributes representing various functions (Stola, 1987), but none of them is synthetic enough to represent itself the particular basic functions of rural areas. Therefore, aiming at minimalisation of the number of attributes on the one hand, and on the other at a most complete and objective representation of the defined functions (and their groupings) we have employed in functional classification of communes in Poland such a set of attributes which would render an elaboration of their measures possible (their methodological and statistical availability). The set includes the following attributes :

1. The proportion of agricultural land (per cent of agricultural land in the total area of Poland);
2. The proportion of socialised agriculture (per cent of socialised agricultural land in the total area of agricultural land);
3. Commercialisation of agriculture (per cent of commercial agricultural production in the total production);
4. The input of direct labour in agriculture (a number of employed persons per 100 ha of agricultural land);
5. Proportion of land under forests (per cent of forests in the total area);
6. The level of the tourist and recreation facilities (a number of beds in tourist and recreation objects per one sq.km.);
7. The level of utilisation of the tourist and recreation facilities (a number of persons availing themselves of the accommodation per one thousand inhabitants);
8. The employment outside agriculture (per cent of the employed persons outside agriculture in the total number of the employed);
9. The employment in industry (and building) according to the place of residence (per cent of the employed persons in industry and building in the total number of the employed); and
10. Commuting to work (per cent of commuters in the total number of the employed).

The first four attributes and their measures pertain, first of all, to the function principal in most regions of Poland, that is, to agriculture. These attributes point either directly or indirectly to the importance of agriculture in the spatial organisation of the countryside, to the comprehensive character of rural economy, including its intensity and links with the market. Moreover, attribute 1 informs indirectly about the degree of utilisation of areas by non-agricultural sectors of the economy. Attribute 5 shows the role of forestry, and the productive and non-productive importance of forests, e.g. the virtual possibilities of development of the recreation function. Attribute 5 in conjunction with attribute 1 informs about the proportion of the so-called other land uses (complement to 100%) in the investigated unit (industrial areas, communication areas, waters, etc.).

Recreational functions (tourism and recreation) are principally represented by attributes 6 and 7. It is obvious that their proper measure could be, for instance, income of the population employed in the tourist services and the share in the national income. However, in Poland, there are no relevant data on the scale of communes. A similar situation occurs when we deal with the choice of attributes representing other functions, first of all, industry. Therefore, of necessity we have limited the number of attributes to those characterising the employment of the population. Attributes 8 and 9 seem to point to a possibly objective role of the non-agricultural functions, primarily of the techno-productive functions. Attribute 8 indirectly shows the size of employment in agriculture (complement to 100%). The development of the dormitory function (residential) understood as a function of the place of residence only and of partial services for the local population but not as a function of the place of work, can be indirectly shown by attribute 10 referring to commuting to work.[1]

Since the character of the individual functions is different, the measures of their role are also different. They are usually coefficients denoted by various denominations, e.g. the percentage, the number of the population or the number of places per a unit of the area. These measures cannot be summed and, therefore, they should be normalised to become comparable. The literature concerned with classification of rural areas lacks such patterns of normalisation of measures and groupings of the examined objects which are obtained by methods rendering the results comparable. Therefore, for these purposes the methods elaborated and employed in typological studies of agriculture have been adopted (Kostrowicki, Szczesny, 1972; Stola, 1977; Kostrowicki, 1980). These methods proved to be the most useful in the studies of multi-attribute objects described by means of different-measure variables (Bielecka et. al., 1977).

The present elaboration is the first, although a simplified attempt at a functional classification of rural areas in Poland on the scale of communes. It was made on the basis of the following measures : the social-occupational structure of the population, commuting to work, landuse, social and property forms prevailing in agriculture, saturation with tourist and recreation facilities, etc. There have been distinguished seven principal classes of communes differing from one to another from the point of view of the leading functions.

1. Highly urbanised or semi-urbanised communes with dominance of non-agricultural functions — the features testified by the proportion of non-agricultural population constituting more than 70 per cent of the total inhabitants. These communes are mostly fund in the urban-industrial agglomerations in Upper Silesia, Bielsko-Biala, Warsaw, Cracow, the Sudeten and Old Polish Industrial Districts etc. The specific character and the level of development of the agglomerations and industrial districts affect the surrounding rural areas which is manifested in different levels of development of the individual functions, in the character of the settlements, ways of living, etc. In addition to industrial functions, agriculture is a source of subsistence too. In some areas like for instance in the suburban areas of Warsaw and Cracow, agriculture is highly specialised and oriented towards the satisfaction of the market. On the other hand, in the areas with a small-scale farm structure of agriculture and a considerable proportion of bi-occupational population like for instance, in the Old Polish Industrial District, in Bielsko-Biala agglomeration, there is the prevalence of semi-subsistence agriculture. In the urbanised areas there develop the so-called agricultural plots (below 0.5 ha) and allotment garden of the non-agricultural population. The agricultural products are destined for self-subsistence of the producers or for the local market. In connection with the growth of pollution of natural environment, also with toxic substances, there has been more and more land which cannot be utilised for food production. In the Katowice voivodship one-half of the total acreage of the agricultural land is unfit for agricultural use and yet it is utilised for production of food.

Concentration of the population in towns leading to demand

of various forms of rest prompts a development of various recreational functions in the areas surrounding the towns and permitting the rest e.g. forests, lakes, rivers, etc. These functions are often combined with building summer houses — the so-called second houses, or summer residences.

Moreover, in the areas with convenient communication connections with urban agglomerations, commuting to work is highly developed, which testifies to the existence of residential functions. For instance, in the vicinity of the Upper Silesian Industrial District, in the Western communes of the Cracow voivodship, the number of commuters amounts to 200 people pre 1000 inhabitants.

2. Another category of communes relatively highly affected by industrial processes and playing various functions are those in which non-agricultural functions offer a source of subsistence for about 50-70 per cent of the rural population. These are mostly mixed areas, agricultural with share of industry and industrial with share of agriculture, often with share of other functions. For instance, the communes with a share of dormitory functions mostly occur in south-eastern Poland, between Opole and Przemysl; with a share of forestry — in the Sudetes and Carpathian regions, in the Lubusz Land, in the Swietokrzyskie mountains, in southern Pomerania, etc. Together with the occurrence of forests there develop the tourist-recreation functions even in the territories where environmental pollution is quite considerable, e.g. the Swietokrzyskie mountains, the Sudety mountains.

   The agricultural functions of the discussed territories show a great differentiation as well as the magnitude and structure of labour input, the size of holdings, the degree of commercialisation, etc., are concerned. In the western part of Poland, particularly in the Leszno, Poznan, Opole, Wroclaw voivodships, the input of direct labour is much lower and the degree of commercialisation higher than the eastern part of Poland, e.g. in the Tarnobrzeg and Przemysl voivodships. This is a result of, above all, differentiation of the level of agriculture development and a differentiation of the agrarian structure.

3. Communes with dominance of agricultural functions measured by the share of agricultural population (above 50%) and by the share of agricultural land (above 50% of the total areas) can be

divided into three principal groups specified by the socio-possessional structure of arable land. The form of ownership of agricultural land — individual or socialised — has a great influence on the management itself as well as on the production effects, the structure and spatial organisation of rural areas.

Among the communes with prevalence of agricultural functions the following categories have been distinguished :

a. the communes with prevalence of socialised agriculture (more than 50% of agricultural land);
b. the communes with a considerable proportion of socialised agriculture (30-50% respectively) and prevalence of individual agriculture (50-70% respectively);
c. the communes with dominance of non-socialised agriculture (more than 70% respectively).

The communes belonging to category 3 foremostly occur in the western and northern territories, for instance, in western Pomerania, Mazurian Lake District, Zulawy region, Lubusz Land and in the Sudety (Fig. 6.1). These communes together with the communes of category 3b cover almost the whole part of western and northern Poland from the Opole voivodship through the Kalisz, Bydgoszcz and Elblag voivodships to the western part of the Suwalki voivodships.

The number of the employed persons in agriculture per 100 ha agricultural land in the communes of category 3a is relatively low — this is due to the relatively low input of direct labour per an area of agricultural land in socialised farms and an insignificant number of small individual holdings. This is particularly true of the northern territories characterised by a low density of population.

The agriculture in the communes of category 3a shows a relatively high input of materialised labour, a high degree of commercialisation and generally, a relatively low agricultural productivity, the individual farming in Opole Silesia, Great Poland and Zulawy region being an exception. The concentration of the agricultural land in the socialised farms has brought about the changes in the configuration of fields, the new arable lands, a liquidation of the boundary strips and woods between the fields, etc., as well as the development of farm building and dwelling houses to fit the needs of large farms. As a result the spatial organisation of the areas utilised by socialised agriculture is different from those owned by individual farmers.

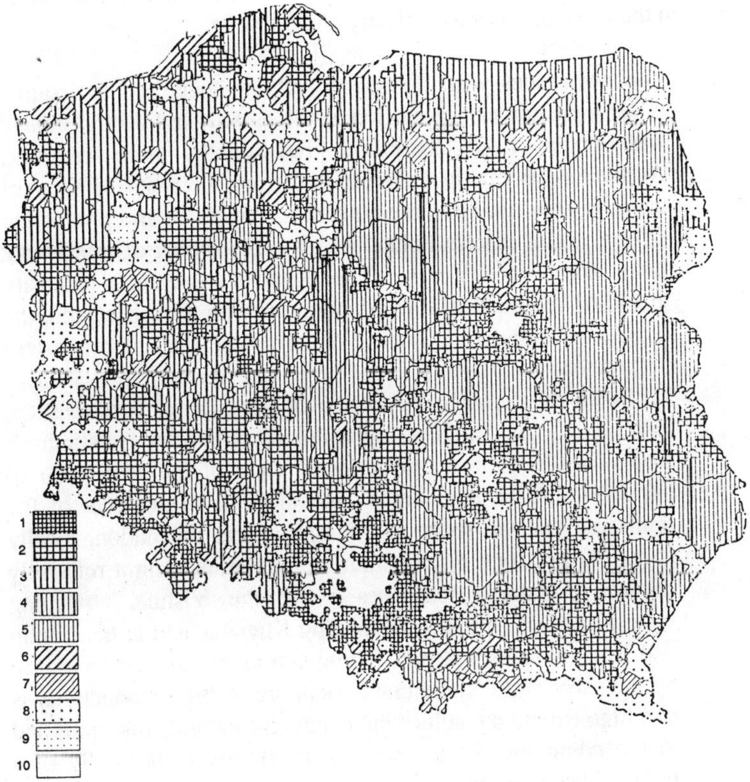

FIG. 6.1: **Functional classification of communes.**
1 — highly urbanised or semi-urbanised, 2 — mixed, agricultural (or industrial) with share of industry (or agriculture), 3 — 5, prevalence of agricultural functions, 3 — prevalence of socialised agriculture, 4 — prevalence of individual agriculture and considerable proportion of socialised agriculture, 5 — dominance of individual agriculture, 6 — 7, tourist-recreational with share of agriculture (6), or forestry (7), 8 — forestry and technoproductive functions, 9 — forestry and agriculture, 10 — towns.

In addition to agriculture, in the lakelands there often occur forestry and the tourist and recreation functions. Non-agricultural population lives on the work in services, industry and building.

4. The communes with dominance of non-socialised agriculture cover a greater part of eastern, south-eastern and central Poland with enclaves of communes functionally more complex. Agriculture is highly differentiated from the point of view of the farm size structure, methods of management and production effects. South-eastern Poland with a high number of small-scale farms has a considerably higher level of agricultural intensity including the input of direct labour, and on the other hand, the degree of commercialisation much lower than north-eastern Poland where medium-size and large-size farms are prevalent.

   In the areas with dominance of agricultural functions the communes of highly agricultural character distinguish themselves both from the point of view of the proportion of agricultural land, sometimes exceeding 70-80 per cent of the total area and the share of the agricultural population, locally above 70 per cent. Such communes are situated in a relatively densely populated territories along the Vistula, stretching widely north of Warsaw, embracing Kuyavia, and in the Lublin and Little Poland Uplends. These communes are characterised by intensive and commercial agriculture. A high productivity is to a high extend a result of the conducive natural conditions and in the communes of the western territories — primarily of a high professional level of the agricultural population. In more urbanised and industrialised areas in south-eastern Poland there is a considerable number of farms owned by bi-occupational population which combines work on the farm and outside it by means of commuting. However, in north-eastern Poland more people emigrate from villages to towns settling there for good, this being due to the lack of possibilities of finding work outside agriculture. There is also a correlation between the territories with dominance of agricultural functions and an intensity of migrations to towns.

5. In the areas with nature attractions such as see side, lake districts, mountains and in the suburban areas slightly affected by industrialisation there develop the tourist and recreation functions. Among the communes disposing of accommodation

for tourists only in one-third of them these functions play a relatively important role (more than 10-50 beds per one sq. km.). Therefore, these communes have been distinguished into the following categories (Fig. 6.1).

(a) Tourist-recreational with share of agriculture,
(b) Tourist-recreational with share of forestry (forests occupying more than 50% of the commune's territory).

The communes in which tourist-recreational functions have developed significantly can be found, above all, at the seaside, in densely forested regions of the Carpathian mountains, in the Sudetes, in the lake lands, especially in the Mazury and Pomerania. Also in some communes lying near the urban agglomerations (Warsaw, Poznan) recreational functions are highly developed.

Apart from the communes already distinguished, which have strongly developed tourist-recreational functions, the communes with areas highly forested (more than 50% of the total areas) can be divided as follows :

6. The communes with a significant share of forestry and technoproductive functions, that is, industry, building.
7. The communes with a considerable share of forestry and agriculture (individual farming in the central, and south-eastern territories or socialised farming in the northern and south-western Poland).

The spatial distribution of these communes is similar to those with tourist-recreational functions, the territories at the seaside being an exception.

This simplified classification of the communes points to a markedly high spatial differentiation of rural areas in Poland from the point of view of the structure of the basic functions. This differentiation grows as we pass from the north-eastern to south-western territories. A relatively great functional differentiation of the communes occurs near the urban-industrial agglomerations. A high degree of functional differentiation in these communes is responsible for the fact that the individual functions, most frequently industry, agriculture and recreation are in conflict with one another, as are the endeavours to preserve the natural environment.

The communes belonging to the above categories differ from one another as far as the developmental level of the basic and secondary (internal) functions are concerned, this depending on the social-economic development, localisation with regard to urban-industrial centers, natural environment conditions, etc.

Spatial differentiation of the functions of communes shows a great need for individual approach while making decisions for its spatial planning

NOTE

1. The adopted list does not contain the attribute directly representing other non-recreational and non-residential functions possibly playing a key role in the functional structure of the seats of the commune authorities, foremostly of the town authorities. This follows, primarily, from the fact that these service functions do not fulfill a condition of commonness. If these functions played a basic role in a commune, on which directly inform attributes 8 and 9, they should be additionally examined to elucidate their character (if they are e.g. communication, educational etc., functions).

REFERENCES

Bielecka K., Paprzycki M., and Piasecki Z., 1977 : Badania nad stosowalnoscia metod ilosciowych w typologii rolnictwa. Refleksje metodologiczne (sum: Research on the applicability of quantiative methods in agricultural typology Methodological reflections), *Przeglad Geograficzny*, 49, 4, pp. 807-17.

Bielecka K., Paprzycki M., and Piasecki Z., 1979 : Ocena stosowalnosci wybranych metod ilosciowych w typologii rolnictwa (sum: Estimation of the applicability of the selected quantitive methods in agricultural typology), *Studia KPZK PAN*, LXVIII, 93. p.

Kostrowicki J., 1980 : Uklad hierarchiczny typow rolnictwa swiata (sum : A hierarchy of world types of agriculture). *Przeglad Geograficzny*, 52, 2, pp. 271-302.

Stasiak A., 1983 : Struktury spoleczno-demograficzne wsi polskiej i jej przemiany (sum: Socio-demographic structures of the Polish countryside and its transformations). *Biuletyn KPZK PAN*, 122, pp. 53-67.

Stola W., 1987 : Klasyfikacja funkcjonalna obszarow wiejskich Polski. Proba metodyczna (sum: Functional classification of rural areas in Poland. An attempt at a methodology). *Ossolineum*, Wroclaw, Prace habilitacyjne series, 166 p.

Szczur M. A., 1983 : Przemysl na wsi polskiej (sum: Industry in the Polish countryside). *Wies Wspolczesna*, 8, pp. 35-46.

7

QUEST FOR LAND REFORMS IN KERALA

M. H. QURESHI AND S. APARNA

INDIA INHERITED a varied and intricate land tenure system after it gained independence. The forms of proprietary rights of land existing in India, viz. State, landlords and peasants, at the time of independence, are related to the forms of feudal land tenures and became more complex during the British rule. In the course of the conquest of India most of her territory was made a component part of the provinces of British India. The proprietary rights on uncultivated land and forests were vested in the State, the bulk of the cultivated land was owned by landlords. Being the "supreme land owners of the country" the British attempted various land revenue systems adopting the forms of land tenure and pattern of cultivation inherited from feudalism. It may be noted that the attempts were made to reform the revenue system obviously geared towards better collection of revenue. By the middle of the 19th century three basic revenue systems had been gradually introduced in the provinces of British India, viz. Zamindari, Ryotwari and Mahalwari, to have a better revenue-collection system.

The zamindari land revenue system entailed the confirmation of the right of ownership of land to individuals, generally from the upper strata of the feudal class. The British accorded legal sanction to the specific local forms of feudal land tenures prevalent in different parts of the country during the late 18th and early 19th century. The tenure holders of various types were, in fact, receivers of rent and during the British rule the transfer of title to a portion of rent and parts of estates took place on a large scale. This led to the emergence of numerous intermediary rent receivers in addition to tenure holders, who had survived from the days of feudalism. The system of receiving rent through landlord-type tenure holders was developed to a smaller extent in Madras and Bombay States but rather to a larger extent in U.P., Bihar, Bengal and Assam.

The proprietary rights of the members of the feudal class as well as of priests were confirmed by the British in the process of creating the *zamindari* land-revenue system. The religious and charitable institutions owned large chunks of good quality land. In the erstwhile Madras province this type of semi-feudal holdings accounted for roughly 10 per cent of all cultivated land in the State (Nair, S.). Some temples had vast land holdings, e.g. the Thiruvilwamala Devaswom, in the former Cochin State, owned 3,500 acres of land, of which 3,000 acres were irrigated land.

The *ryotwari* land revenue system had no intermediaries between the State and the farmer. This brought the State and the tiller in direct relationship unlike the zamindari system where intermediaries formed an important segment in the chain of relationship. It gave the ownership rights on land to the members of village community. But this system also witnessed the emergence of landlords belonging to the upper social strata of the village community and money-lenders, traders and professionals virtually became absentee landlords. Hence the burden of the revenue was shifted to the actual tiller of the soil.

The *mahalwari* system, as prevalent in Punjab and adjoining States, envisaged communal collection of land revenue. The whole village community or a number of villages were taken together for revenue assessment and the revenue was collected by revenue officials through the village Lambardar (Numberdar). The institution of Lambardar still exists in the villages of Haryana and a few other parts of the country.

The land reforms introduced in the post-independence period beginning with the abolition of zamindari ushered in a new era of land relations in the countryside. While the large estates were broken by these reforms, the later laws, however, could not break up the land. Concentrations in the hands of influential landholders became common. It is a widely accepted fact that the legal loopholes and lack of enthusiasm and political will at the level of implementation made the various laws ineffective instruments.

Experience in Kerala

The main objective of this paper is to examine the changes in the land relations that have taken place due to Land Reform Acts in Kerala during 1850-1971. The State of Kerala came into being in November,

1956 as a result of the recommendations of the States Reorganisation Commission. It includes all the Malayalam-speaking regions, i.e., the erstwhile State of Travancore-Cochin (excluding four Tamil-speaking taluks), Malabar district of the former Madras province and Kasargoda taluk of South Canara district. The State of Travancore-Cochin was formed in 1949 after the integration of the princely States of Travancore and Cochin which were separate entities before independence. Thus, in effect, Kerala was formed by combining three separate political units. Wedged between the Arabian Sea and the Western Ghats in the south-western corner of the peninsula, Kerala sprawls over three physiographic distinct zones. The coastal low lands, the middle zone of medium elevation and the mountainous zone with higher elevation which has uniformly high temperatures and receives heavy rainfall. These zones extend across all the erstwhile political units.

An Overview of Land Reform Measures

To have an insight into the land reform measures adopted in the erstwhile political units one has to look back into the historical past particularly in the recent past, and it has been essayed to scan through the measures taken after 1850 in the three constituent units separately.

Land Reform Measures in Travancore

The land in Travancore before independence was either under Government ownership (*Pandarvaka*) or was owned by landlords (*Jenmies*). The Jenmies had created various types of tenancies and extracted revenue in cash, kind or even labour. The tenant could be evicted and divested of land at will. Travancore State initiated land reforms by issuing the Pattom Proclamation, in 1865. The Act provided for the fixity of tenure of the tenants on the Pandarvaka land. It also conferred proprietary rights on some types of tenants. Two years later, in 1867, another progressive measure was taken which was known as Jenmi Proclamation. This proclamation could not provide ownership rights to the tenants but it did save them from being evicted provided they paid the rent and other customary dues. This Act was reinforced and made more effective by Jenmi and Kudiyan Regulation of 1895-96 which not only clearly defined the rights of Jenmies and Kudiyans but also provided the Kudiyans with permanency in occupancy rights. The Act was again amended in 1932-33 which converted all the Kanom tenants of Jenmom land into proprietors of such lands. The Kudiyans were under obligation to pay only the *Jenmikkaram* to the State. The State was responsible to remit it to the Jenmies.

The Travancore Land Tax Proclamation of 1946 was more comprehensive. It brought almost all the assessed lands under a uniform rate of assessment and removed the revenue-tenurial differences. Owing to the uniform assessment of levy on all the lands, the owner had the incentive to put all land under some use or the other.

But despite these various legislative measures facilitating the development of agriculture on the basis of "peasant proprietorship", leasing of land and tenancy problems created thereby increased during the whole period.

Land Reform Measures in Kochin

The settlement Proclamation of 1904-05 conferred the ownership rights on *Pandaravaka* tenants. The tenants of incorporated *devaswom* got the fixity of tenure by paying a higher rate of land revenue through a Devaswom Proclamation in 1909.

The Cochin Tenancy Act of 1914-15 provided for the award of compensation for improvements of land by tenants. The 1938 Cochin Tenancy Act broadened by definition of *Kanom* tenants to include a few more categories of tenancy. The Cochin *Verumpattomdars* Act, 1943 granted permanent occupancy rights to Verumpattom tenants irrespective of the nature of their tenancy and the duration of occupation of land. It also provided the tenants to pay a contract rent. The Devaswom Verumpattom Proclamation of 1943 besides conferring permanent occupancy rights on all tenants fixed their rents also.

Absentee landlordism continued to prevail in extreme form on Janmom land. The approval of the "contract rent" restricted the bargaining power of the enterprising tenants in leasing in the land on moderate rents and their margin of profit from cultivation. With the lessees and sub-lessees securing fixity of tenure, the *Kanom* leases and other intermediary tenancies became unpopular. The land-owners preferred to mortgage land when they needed lump-sum money, because they did not, in that case, have to forgo their occupancy rights for ever. Thus, in place of numerous types of leases in practice earlier, there emerged tenancies-at-will with characteristics of share-cropping and mortgaging of land.

Land Reform Measures Taken After the Creation of Travancore-Cochin State in 1949

In 1949 immediately after the integration of two States — Travancore and Cochin — two important legislations were enacted. Firstly, the Kandu Krishi Proclamation of 1949 transferring the property of Maharaja of Travancore to the Government. Secondly, the Holdings (Stay of Execution Proceedings) Ordinance of 1949, a temporary measure, intended to stay the proceedings in the execution of the decrees of eviction against ordinary tenants, in possession of land for a specified period, pending enactment of a comprehensive legislation, which came in 1950. Five legislations were enacted during 1955-56 in order to bring a parity between the two constituents, i.e. Travancore and Cochin. The Travancore-Cochin Prevention of Eviction of Kudikiddapukars Act, 1955, the Travancore-Cochin Land-Tax Act, 1955, the Kanom Tenancy Act, 1955, and, the Travancore-Cochin Compensation for Tenant's Improvement Act, 1956, were intended to introduce uniformity in the legislative provisions in Travancore and Cochin. The first Act unified the provisions contained in the Travancore Prevention of Eviction Act, 1949 and the Kochin Proclamation of 1947 into one legislation prohibiting the eviction of *Kudikiddapukars* from their huts. The Land Tax Act extended the basic tax system to Cochin and brought all the assessed land in the State under a uniform land revenue assessment. The Kanom Tenancy Act was intended to bring the *Kanom* land in Cochin under the same provisions as there of the Jenmi-Kudiyan regulation of Travancore. The compensation Act introduced uniform law for the whole State ensuring fair compensation for improvements made by the tenants at the time of their eviction. The fifth Act known as the Edavagai Acquisition Act, 1955 abolished the free-holds enjoyed by four chiefs of Travancore, viz. Poojar, Edapally, Kilimanoor and Vanjipuzha, after paying them compensation. After the enactment of this Act every registered land-holder of Edavagai land was subject to payment of basic tax to the Government only. The tenants who held land under the chiefs either from the home-lands or the waste-lands, however, did not benefit, in any way, from this Act.

Land Reform Measures in Malabar during the Pre-Independence Period

The first step towards tenancy reform in the district was the enactment of the Malabar Compensation for Tenant's Improvement Act of 1887 which was subsequently amended in 1900. This Act intended to control, firstly, the arbitrary eviction and, secondly, rack-rending. The curbs brought about by the provisions of the Malabar Tenancy Act of 1929 were beneficial to the interests of the actual cultivators. It guaranteed fixity of tenure and fair rent to all cultivating *Verumpattomdars*. It also conferred the rights of renewal on all customary Verumpattomdars, Kuzhikanomdars and certain Kanomdars. The Act did not confer security of tenure on non-cultivating *Verumpattomdars*. It could safeguard the interests of the upper strata of tenants, i.e. the *Kanoms* and the customary *Verumpattomdars*.

Land Reform Measures in Malabar during the Post-Independence Period

The Malabar Tenancy Act of 1929 was amended during 1951 and also in 1954. The provision of renewal of tenancy to various categories of tenants in the 1929 Act was abolished and all the categories of tenants covered by the 1929 Act, viz. *Kanoms, Kuzhikanoms*, etc., got fixity of tenures. All the holders and tenants of Kudiyiruppu were conferred heritable occupancy rights. Fair rents were reduced for Verumpattomdars and their liability to pay advance rent was also done away with. This Act further made provision for eviction of the tenant, in case the landlord wanted the land for self-cultivation or for construction of a building for his own residence. This provided a major loophole to effect eviction which was modified by the amendment of 1954. It made provision to protect all the cultivating Verumpattomdars from being evicted by the landlord on the plea of self-cultivation provided the tenant has occupied the holding for a period of six years or more. The occupants of less than six years could not be saved by the amendment.

Land Reform Measures after the Formation of Kerala State in 1956

The major task after the formation of the State was to remove the dissimilarities in the legislative provisions between the various constituent units. The Communist Party came to power and soon after important Acts were passed, viz; (i) the Kerala Stay of Eviction Proceeding Act, (ii) the Kerala Land Tax Act, (iii) the Kerala Land Relinquishment Act. The first Act stayed the eviction proceedings against all categories of tenants, lessees and Kudikidappukars; the second Act extended the basic tax system of assessment to Malabar and the third Act was intended to check the encroachment on Government land. A fourth Act legalised the relinquishment of lands by owners in favour of the Government.

The Kerala Agrarian Relations Bill 1960 conferred fixity of tenure on all types of tenants including all earlier categories, new types of tenancies and the landless Kudikidappukars. The Government also brought forward the Janmikkaram Payment Abolition Bill in order to abolish the recurring and periodical payment of *Janmikkaram* to Janmies under the Travancore Janmi and Kudiyan Act of 1896 (amended in 1932). It provided for payment of compensation to and conferment of full ownership rights on their holdings to the tenants. The Act, when passed, made a distinction between the mode of payment of compensation to religious and charitable institutions of public nature and other Janmies.

The Kerala Land Reforms Act, 1963 aimed of bringing effective unification of various types of land relations existing in the State. It provided for higher rent rates but at the same time granted security of tenure for all tenant subject to a limited right of resumption by owners for self-cultivation or for extension of the places of worship or for construction of residential buildings. It also provided for a ceiling on existing holdings and on further acquisition. Under the Act the rates of fair rent were made uniform throughout the State and small holders were granted special benefits, viz. security of tenure, fixation of fair rent and the right to purchase the landlord's rights, and become full owners of the land. Land tribunals were established to safeguard the rights of tenants. While one appreciates the intentions of these Acts, the implementation by various Governments has been given a lukewarm treatment except for a few provisions which could be implemented in letter and in spirit.

The Pattern of Land Distribution

The impact of the various land reform measures taken overtime should get reflected in the pattern of land distribution and can be roughly measured by looking at the concentration of land in various size classes of holdings. In the following paragraphs an attempt has been made to examine the pattern of land distribution in the State as well as to examine its spatial pattern (districtwise) Fig. 7.1, 7.2 and 7.3. The exercise is based on the data obtained from *The Agricultural Census* of Kerala, 1971. This attempt has been made by looking at the Ginni's coefficient separately for:

(i) total holdings,
(ii) wholly owned and self-operated holding, and
(iii) partly owned and partly rented holdings.

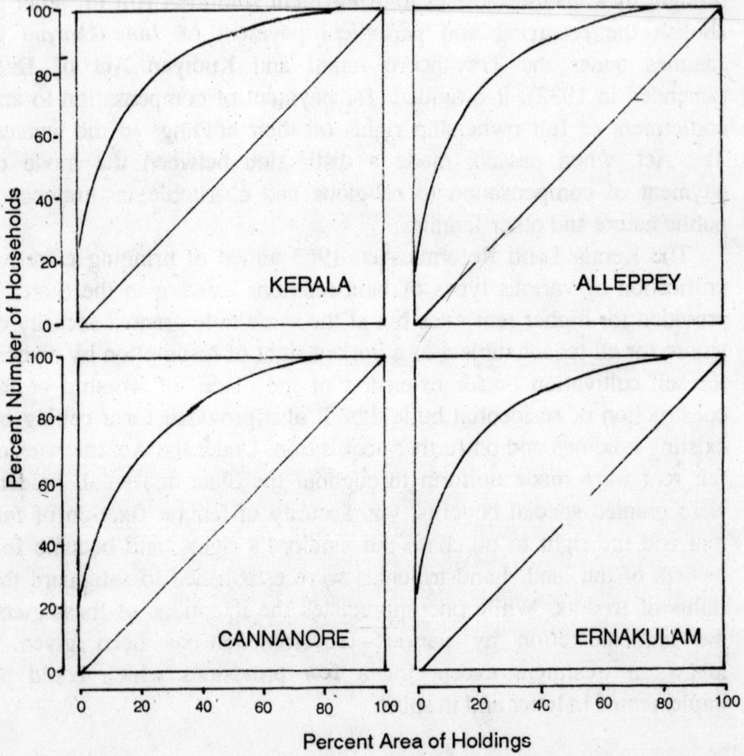

FIG. 7.1: Districtwise pattern of land distribution in Kerala, 1971.

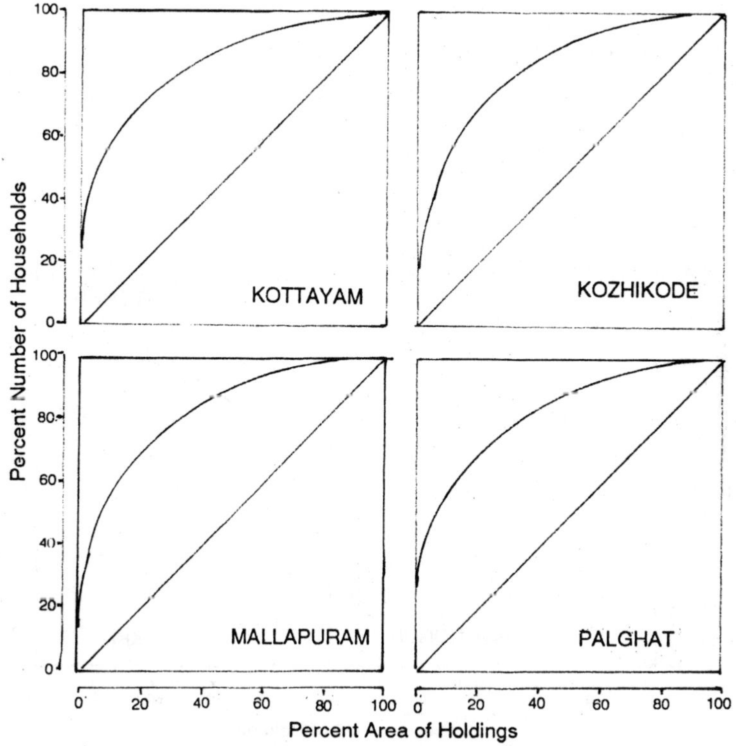

FIG. 7.2: Districtwise pattern of land distribution in Kerala, 1971.

In order to depict the spatial pattern Lorenz curves have been drawn for individual districts as well as for the State. Ginni's coefficients also provide a measure to understand the spatial pattern of this phenomenon. Looking at the total holdings we find that the average size of a holding for the State of Kerala is 0.57 hectares. Table 7.1 reveals that the holdings show a distribution pattern which is inclined towards concentration of land in the small-size classes.

It is found that 18 per cent of the area is distributed over 73 per cent of households. Most of the districts conform to the State pattern in the land distribution. Idikki district shows the highest concentration in the State where Ginni's coefficient is 0.72. This also indicates that the holding sizes are mostly small and medium and there are only very few large holdings. In the State as a whole 78.21 per cent of the area under

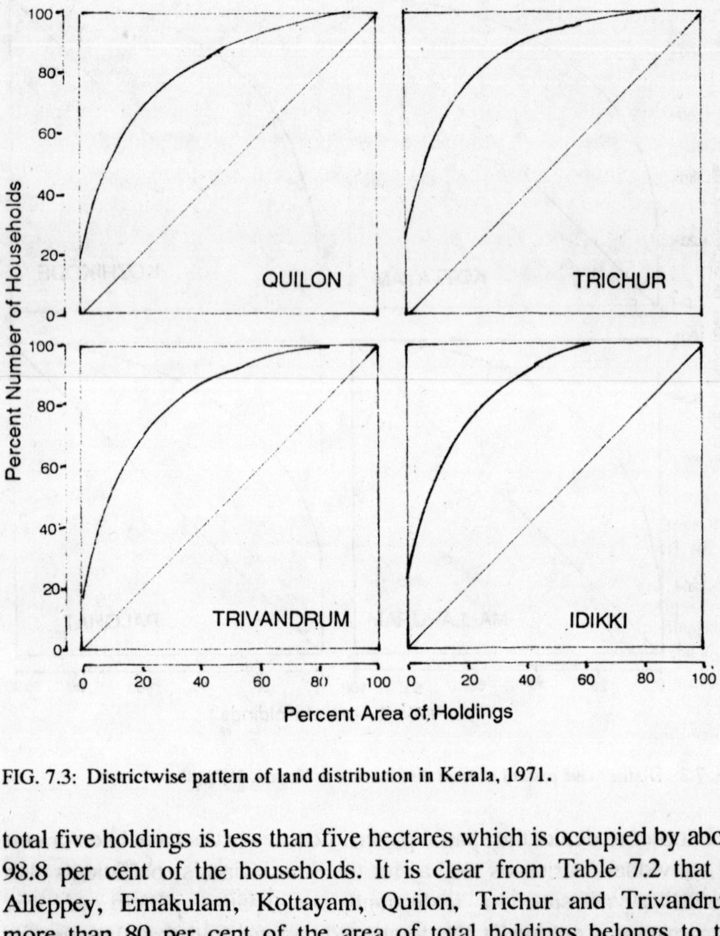

FIG. 7.3: Districtwise pattern of land distribution in Kerala, 1971.

total five holdings is less than five hectares which is occupied by about 98.8 per cent of the households. It is clear from Table 7.2 that in Alleppey, Ernakulam, Kottayam, Quilon, Trichur and Trivandrum more than 80 per cent of the area of total holdings belongs to the holding size of less than five hectares. In Cannanore, Kozhikode, Mallapuram and Palghat, it is between 70 and 80 per cent while only in Idikki it is 50 per cent. Area under large holdings is limited and varies from 0.9 per cent in Kottayam to 6.6 per cent in Palghat except Idikki and Cannanore where 39.72 and 16.4 per cent of the area of the total holdings respectively is found in more than 50-hectare size class. The table further reveals that the concentration tends to exist with the size classes of less than five hectares rather than in large-size classes. The pattern leads to believe that there has been a breakthrough in land distribution as a result of various measures adopted by the erstwhile constituents and the State of Kerala after its formation. Table 7.2 and Table 7.3 depict the distribution pattern of the households and area under the households in various size classes. The majority of the households fall in the size class of less than five hectares. Nowhere in Kerala, are there more than 0.50 per cent of the households which fall in the size class of more than 50 hectares. Cannanore and Idikki are the only two districts where 0.02 and 0.35 per cent of the households have 16.40 and 39.72 per cent of the area respectively in the size class of more than 50 hectares.

TABLE 7.1: Ginni's coefficients (table holdings no. and area)

Districts	Total holdings	Partly owned and operated	Partly owned and partly operated
Alleppey	0.64	0.57	0.36
Cannanore	0.66	0.62	0.40
Ernakulam	0.68	0.63	0.48
Kottayam	0.66	0.59	0.49
Kozhikode	0.65	0.43	0.26
Malappuram	0.66	0.69	0.44
Palghat	0.65	0.59	0.47
Quilon	0.61	0.54	0.43
Trichur	0.68	0.62	0.44
Trivandrum	0.65	0.55	0.59
Idikki	0.72	0.64	0.84
Kerala	0.69	0.63	0.80

It should be clearly understood that land reform is not an end in itself, rather it is a means for better utilisation of resources and removing the income disparities in the rural areas. The success in this direction depends not in the enactment of a host of Acts and laws but in the implementation. We are aware of the fact that no legal provision is perfect and free from loopholes under the existing social set-up. The real impact of such legislations on land distribution and the regional variation in this respect can be understood by a microlevel survey reflecting the social dynamism in the countryside of Kerala.

TABLE 7.2: Proportion of area in each holding size class

Size Class	Kerala	Alleppey	Canna-nore	Erna-kulam	Kottayam	Kozhi-kode	Malla-puram	Palghat	Quilon	Trichur	Trivan-drum	Idikki
1 hectare	18.34	32.55	12.37	20.28	13.83	16.68	17.34	7.21	34.01	24.41	35.59	4.50
1 to 5 hectare	59.87	58.14	58.34	63.26	71.68	62.45	59.0	65.92	55.53	62.62	52.45	45.54
5—10 hectare	3.3	2.44	3.11	3.69	3.76	4.99	4.48	5.13	1.39	2.61	1.15	2.25
10—20 hectare	5.97	3.05	6.12	7.29	5.00	7.42	7.19	12.34	2.26	4.98	2.21	3.99
20—30 hectare	2.39	1.22	2.52	2.44	1.93	2.43	3.61	4.40	1.11	1.80	1.54	1.98
30—40 hectare	0.86	0.46	0.82	0.41	1.23	0.71	1.54	1.37	0.43	0.61	0.34	1.21
40—50 hectare	0.41	0.32	0.32	0.23	0.66	0.76	0.20	0.74	—	0.72	0.32	0.20
50 and above	8.83	1.82	16.40	2.37	0.91	4.56	6.56	2.89	5.27	2.25	6.40	39.72

Computed from : *The Third Decennial Census of Agriculture, 1970-71, Report on Kerala State,* Vol. II, Statistical Tables, Bureau of Economics and Statistics, Govt. of Kerala, Trivandrum 1973.

TABLE 7.3: Proportion of nos. of households each size class

Size Class	Kerala	Alleppey	Canna-nore	Erna-kulam	Kottayam	Kozhi-kode	Malla-puram	Palghat	Quilon	Trichur	Trivan-drum	Idikki
1 hectare	71.82	82.79	58.09	75.05	67.03	66.08	67.44	52.85	80.21	79.35	83.72	44.51
1 to 5 hectare	26.98	16.83	40.22	23.91	31.86	32.24	30.76	43.53	19.49	20.04	16.04	53.18
5—10 hectare	0.43	0.18	0.63	0.39	0.50	0.76	0.56	1.21	0.11	0.23	0.08	0.83
10—20 hectare	0.52	0.16	0.82	0.54	0.47	0.75	0.71	1.96	0.14	0.30	0.11	0.87
20—30 hectare	0.10	0.03	0.18	0.09	0.09	0.12	0.18	0.34	0.03	0.05	0.04	0.25
30—40 hectare	0.02	0.01	0.03	0.01	0.03	0.02	0.05	0.06	0.01	0.01	0.00	0.08
40—50 hectare	0.01	0.00	0.01	0.00	0.01	0.01	0.01	0.02	0.00	0.01	0.00	0.01
50 and above	0.02	0.00	0.02	0.01	0.01	0.02	0.02	0.03	0.01	0.01	0.00	0.01

Computed from : *The Third Decennial Census of Agriculture, 1970-71, Report on Kerala State,* Vol. II, Statistical Tables, Bureau of Economics and Statistics, Govt. of Kerala, Trivandrum 1973.

GLOSSARY

Devaswom : Belonging to the deity.

Edavagai : Freehold estates owned by the four former chieftains of Travancore, viz. Poonjar, Edapally, Kilimanoor and Vanjipuzha.

Jenmi : Upper-caste land owners mostly Brahmins

Janmikkaram : Commuted value of the rent fixed by the State.

Kanom : A customary tenure interpreted by the British courts a possessing the character of both lease and mortgage.

Kanomdar : Holder of the Kanom tenure.

Kudikidappukar : One who does not have homestead or a land on which to construct one and has been given possession for this purpose.

Kudiyan : Usually a holder of Jenmikkaram, land.

Kuzhikanom : Lease of wasteland in favour of a tenant who has to make improvement on land.

Pandarvaka : Belonging to the State.

Patta : The title deed of ryotwari land issued to the peasant by the State.

Pattadar : Holder of Patta.

Verupattom : Tenancy at will.

Verumpattomdars : Tenants leasing in land on tenancy at will.

REFERENCES

Agarwal, S. K., 1971 : *Economics of Land Consolidation in India*, S. Chand & Co. (Pvt.) Ltd.

Appu, P. S., 1971 : *Ceiling on Agricultural Holdings*, Ministry of Agricultural, Govt. of India.

Baden Powell, B. H., 1892 : *Land System of British India*, Vol. III, Oxford, London.

Dantwala M. L. and Shah, 1971 : *Evaluation of Land Reforms*, Department of Economics, University of Bombay.

Graham Hallett, 1960 : *The Economics of Agricultural Land Tenure*, handbooks.

Gregory Kotovsky, 1964 : *Agrarian Reforms in India*, People's Publishing House.

Joshi, P. C., 1975 : *Land Reforms in India*, Institute of Economic Growth.

Khusro, A. M., 1973 : *The Economics of Land Reform and Farm Size in India*, Macmillan India.

Malviya, H. D., 1954 : *Land Reforms in India*, Economic and Political Research Department, All India Congress Committee, New Delhi.

Mukherji, Karuna, 1952 : *Land Reforms*, H. Chatterjee & Co., Calcutta. Nair, S., *Land Reforms in India*.

Patel, G. D., 1957 : *The Land Problem of Recognised Bombay State*, N. M. Tripathi (Pvt.) Ltd.

Pillai, V. R. and P. G. K. Panikar, 1965 : *Land Proclamation in Kerala*, Asia Publishing House.

Sharma, R. S., 1971 : *Land Revenue in India—Historical Studies*, Motilal Banarsi Dass.

Thorner, Daniel, 1976 : *Agrarian Prospect in India*, Allied Publishers.

Varghese, T. C., 1970 : *Agrarian Change and Economic Consequences*, Allied Publishers.

Chief Minister's *Conference on Land Reforms*, 1972 : *Summary record*, Govt. of India, New Delhi, 23rd July.

Planning Commission, Govt. of India, 1958 : *Reports of the Committees of the Panel on Land Reforms;* Planning Commission.

————, 1963 : *Progress of Land Reform*, Manager of Publications.

Report of the Land Reforms, 1968 : *Survey in Kerala, 1966-67*, Bureau of Economics and Statistics, Trivandrum.

8

CASTE AND HIERARCHY OF PRIMARY OCCUPATIONS IN INDIA
A Geographical Analysis

NOOR MOHAMMAD

CASTE AND PRIMARY occupations are intrinsically linked with each other in India since the dawn of history. In the past every caste had occupation exclusively of its own. All civilised societies and even the primitive and non-literate ones had different view points regarding the occupations adopted by them. It was traditional and hereditary occupation of its members and to abondon it in search of others was at least not proper, if not actually sinful. But with the passage of time, the nature, structure and composition of caste is changing due to cross breedings, inter-caste and inter-religion marriages, and so is the case with the occupational structure which is becoming more and more complex day by day due to technological changes in almost all walks of life. In such a caste-occupational dynamism caste, occupation relationship still exists together with a high degree of hierarchy of occupation at micro level to a certain extent in general and to a large extent in particular. The degree of relationship depends, by and large, on level of technological development and modernisation. In the backward areas dairy farming, crop farming, vegetable farming, poultry farming, fishing and hunting and gathering are still associated with certain castes may be in the modified form.

However, with the modernisation and technological development which is percolating gradually in rural areas, occupational structure is becoming complex day by day. People do not get satisfied with a single occupation and in several cases it does not fulfil the necessities of the household as well. It is more so when the family is large and the demands are varied. Therefore, not only the head of the household but also the other members of the family go for several subsidiary occupations which contribute to the family income. On the basis of importance — social and financial, the occupations may be divided into different hierarchical orders which would be helpful in planning the overall socio-economic development of the region. Since the country is too large, it will not be possible to study this problem based on primary data for the country as a whole, the state of Uttar Pradesh has been chosen in the present study.

The Geographical Profile of the Area

The State of Uttar Pradesh, situated in the northern part of India between 23° 52′ — 31° 28′ N. Lat. and 77° 04′ — 84° 38′ E. Long. covers an area of 294, 362 square kilometres. It had 48 districts in 1931 which have now been split into 56 districts (Fig. 8.1). Lying in the Ganga plain, the entire state, with few exceptions, is a vast homogeneous plain mainly drained by the rivers Ganga, Yamuna, Ghaghara, Gomati Rapti etc. The state enjoys sub-humid climate with three well marked seasons: Summer, rain and winter. The soil is fertile alluvium deposited by the river channels. It has a predominantly agrarian structure. Agriculture, engaging roughly 80 per cent of the total population directly and indirectly, is the main occupation of the people and it provides more than 50 per cent of the state income. The state is not favourably endowed with mineral resources and its industrial base is also comparatively poor. It has a fairly good network of transport and communication, although there are a lot of regional variations in it. Uttar Pradesh is one of the densely populated states of the country and has a very complex population structure. It is more so in case of caste composition of population. More than hundred castes are found here but the major ones are Brahmin, Rajput, Scheduled caste, Muslim, Jat, Kurmi etc. Since the state is at comparatively lower web of the level of technological advancement, the hierarchical association between caste and the primary occupation is more glaring and strong but it has not been properly studied so far.

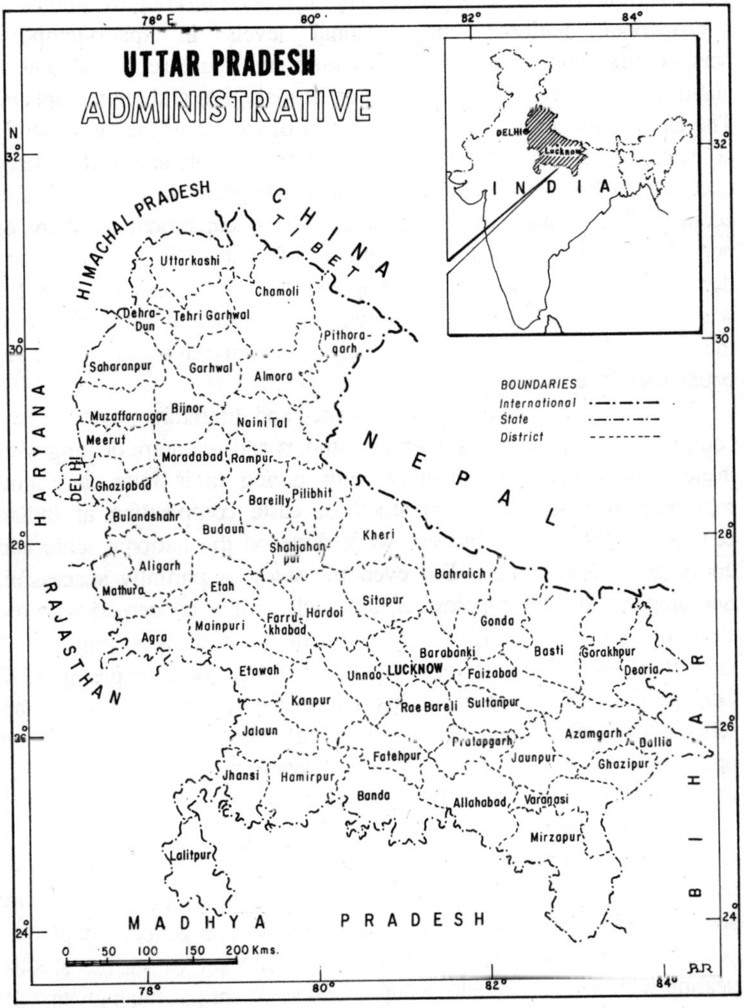

FIG. 8.1: Location of study area.

Review of Literature

The castes and its various parameters have been thoroughly studied at international, national and regional levels in spatio-temporal perspectives. The caste system, its nature and structure, has also been elaborately studied by sociologists, anthropologists and ethnographers. The caste class and power have been one of the major themes of study of sociologists. Similarly various primary, secondary and tertiary occupations have been studied by the geographers in particular, and economists and historians in general. The occupational structure of population has also been a pivot of study for geographers, demographers economists, sociologists, and historians all over the world (Berry, et.al., 1976). Enumeration of occupation has been one of the major themes of population census enumerations in almost all countries of the world.

Thus there are plenty of documents and literature on these two components. But there is a dearth of studies and literature dealing with these two components together with giving their inter and intra relationships Risley's (1915) study of caste composition at Indian census of 1901, is the pioneer work done on the national scale and under government aegis. But even his study is partially successful. Sociologists and anthropologists, although mainly concerned with the study of caste class and power, have done some study in this respect from their own points of view (Omvedt, 1980). For example, in 1945 Gadgil in the first report of his Poona Survey, has published his scheme of occupational (socio-economic) categories based on their nature and status. His study of principal weaving castes of Maharashtra and the changing nature of their occupation, is one of the basic works of caste-occupation relationship. Ghurye (1961) has written a voluminous book on *Caste, Class and Occupation* which sounds to be reasonably dealing with caste and occupation. But it gives more specific treatment to caste and its correlates with a general chapter on occupation and caste. In this chapter too, a review of various studies dealing with occupation and occupational structure has been given with special mention of few studies which deal with caste and occupation at micro or regional level.

The geographers did study the primary occupations especially crop farming, dairy farming, vegetable farming etc. but they were busy in analysing the spatial distribution of agriculture with respect to pysio-socio-economic conditions (Singh, 1970, Mohammad, 1978, Symon, 1978), and the institutional base of agriculture especially caste has not been studied in proper perspectives. The studies which are available are sketchy and too general in this respect (Buchanan, 1959, Symon, 1978, Mohammad 1981). Thus in geographical literature the study of caste and primary occupations with their hierarchy is almost non-existent as far as the author knows. The author has given the Field Survey Projects on this issue to his M.A. Final year students. The surveys have been conducted in the villages of the Union territory of Delhi and it gives different kinds of association and hierarchical orders of occupations which has really prompted the author to conduct a thorough study of the problem.

Research Design and Methodology

The study is essentially based on primary data but secondary data have also been used for background study and information. Stratified purposive sampling technique has been used in collection of the primary data. The state has been divided into caste-combination regions on Doi's method using 1931 data for caste — the only published or otherwise data available on caste at district level (Fig. 8.2). One village has been selected from each caste-combination region except one from where two villages have been selected due to the large size of the region. From these selected villages 1033 respondents, belonging to different castes, have been chosen who have been thoroughly interviewed according to a well-structured questionnaire, which beside containing a number of quarries, focus on six primary occupations viz. crop farming, dairy farming, fishing, vegetable farming, poultry farming, and piggeries (Table 8.1). It also includes question regarding the occupations such as labour, service, carpentry, weaving etc., being adopted by the family members. Questions have also been made in order to ascertain the hierarchy of occupation.

These questionnaires have been processed and on the basis of the response of the respondents and the over all contribution of different occupations, the First, Second and Third occupations have been ascertained. Finally an attempt has been made to examine and explain caste-wise hierarchy of occupation.

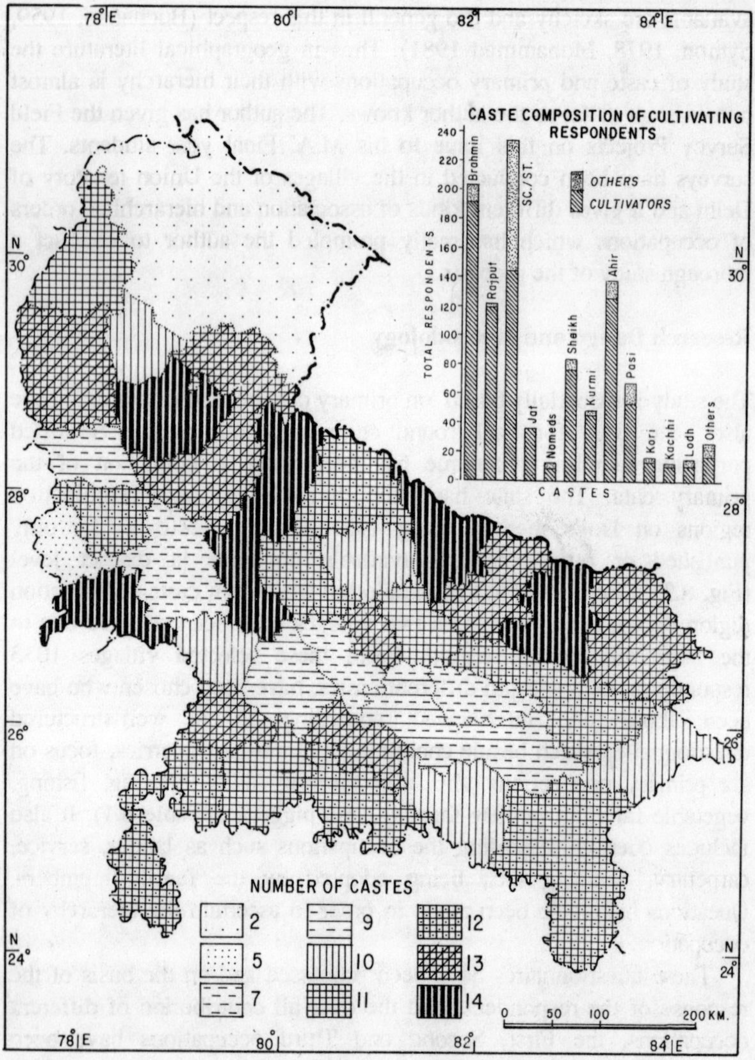

FIG. 8.2: Caste-combination region.

TABLE 8.1: Caste-wise number of selected respondents

Castes	No. of respondents	% age to total
Brahmins	204	19.76
Rajput	123	11.91
Scheduled castes	235	22.86
Jat	38	3.68
Nomads	14	1.35
Shaikh	85	8.04
Kurmi	49	4.74
Ahir	139	13.47
Pasi	69	6.69
Kachhi	14	1.36
Lodh	17	1.68
Kori	18	1.75
Other castes	28	2.71
Total	1033	100.00

Castes and Their First Occupations

Crop Farming

A caste-wise summary of the respondents with their first occupation is given in Table 8.1 which shows that about 57 per cent respondents have crop farming as first occupation followed by labour which accounts about 13 per cent of total respondents. In crop farming Brahmins dominate the scene by contributing about 80 per cent of total respondents having crop farming as their first occupation. More than three-fourth of Brahmin respondents are basically cultivators. Rajput is next caste, among which about 78 per cent respondents have crop farming as their first occupation (Table 8.2). Ninety six Rajputs, accounting to 16.22 per cent of the respondents having crop farming as first occupation, have it their first occupation. Crop farming is first occupation among 29 per cent of scheduled caste respondents, and they constitute about 12 per cent of respondents under this category. Shaikh is an other caste with 67 per cent of its respondents having crop farming as their first occupation. They account about one-tenth of respondents under this category. Fifty Ahirs, accounting 36 per cent of total Ahir respondents, have crop farming as their major occupation; and they account about 8 per cent of the respondents under this group.

About 55 per cent Pasi have crop farming as their main source of livelihood, whereas about 80 per cent Kurmi farmers depend on crop farming. Both account 6.42 and 6.59 per cent of total respondents of this category respectively. Jat is an important caste with crop farming and dairy farming as their main source of livelihood. The former engages about 95 per cent of Jat respondents, while the latter only 5 per cent. Among Kachhi and Lodh about 58 per cent of total respondents depend mainly on crop farming and they together account about 3 per cent of the respondents under this group. Gujar, Gadariya and Banjara, although account only 1.35 per cent of total respondents, but 57 per cent respondents of these castes have crop farming as their first occupation (Fig. 8.3).

TABLE 8.2: Castes and their first occupations (number of respondents)

Major Castes	Crop Farming	Dairy Farming	Fishing	Vege-table Farming	Poultry Farming	Pigge-ries	Service	Labour	Others
Brahmin	162	5	—	8	—	—	20	8	1
Rajput	96	5	—	2	—	—	16	4	—
S.C./S.T.	69	9	2	—	—	72	10	70	3
Jat	36	2	—	—	—	—	—	—	—
Nomads	8	6	—	—	—	—	—	—	—
Shaikh	57	8	—	1	6	—	2	9	2
Kurmi	39	3	—	—	—	—	3	2	2
Ahir	50	75	—	3	—	—	3	8	—
Pasi	38	—	—	13	—	1	—	15	2
Kachhi	8	—	—	1	—	—	1	4	—
Lodh	10	—	2	—	—	—	—	5	—
Kori	4	1	—	9	—	—	—	3	1
Other Castes	15	4	1	—	—	—	1	3	4
Total	592	118	5	37	6	73	56	131	15

Dairy Farming

It is next important type of farming being adopted by more than 11 per cent of the total respondents. Animals, specially cows, are kept by Hindus on account of religious feelings as well. Ahir is most important caste in this respect accounting about 64 per cent of the total respondents having dairy farming as first occupation. More than 54 per cent Ahir respondents have dairy farming as their first occupation.

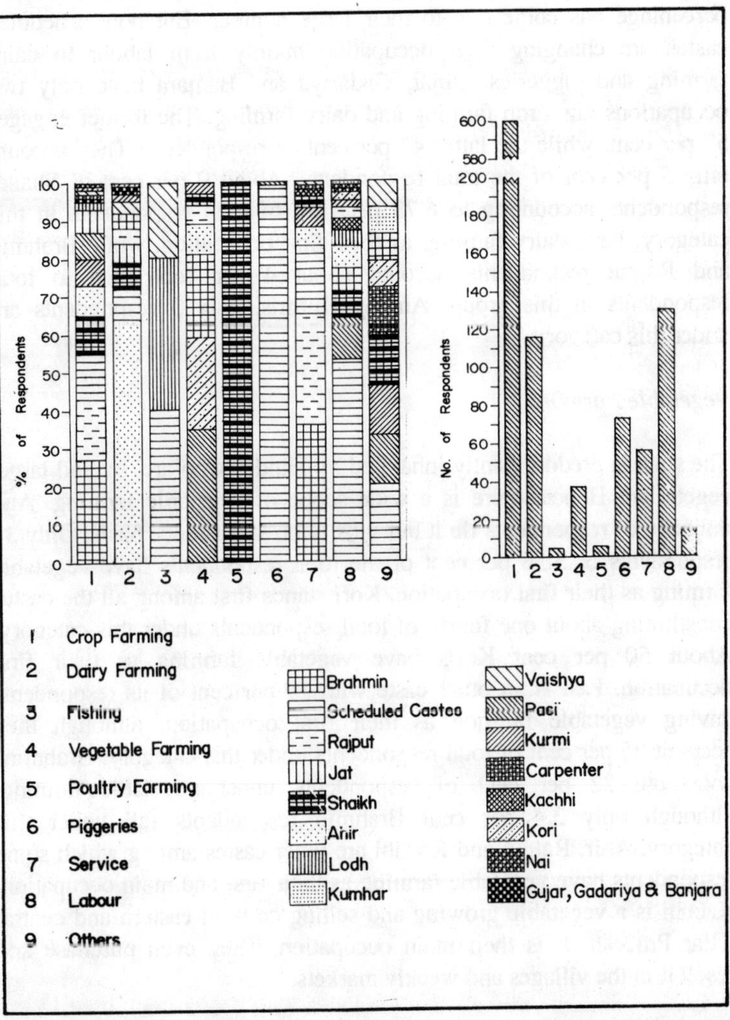

FIG 8.3: Major occupations (I occupation).

Scheduled Caste rank second contributing about 7 per cent of total respondents under this group but only 3.8 per cent of scheduled caste respondents have dairy farming as their main occupation. The higher percentage has come due to their large number. But now scheduled castes are changing their occupation mainly from labour to dairy farming and piggeries. Gujar, Gadariya and Banjara have only two occupations viz. crop farming and dairy farming. The former engages 57 per cent, while the latter 43 per cent of respondents. They account only 5 per cent of the total respondents. About 9 per cent of Shaikh respondents accounting to 6.78 per cent of total respondents in this category, have dairy farming as their first occupation, while Brahmin and Rajput respondents account about 4 per cent each to total respondents in this group. Among Kurmis, only 3 respondents are under this category.

Vegetable Farming

The state is predominantly inhabited by Hindus who are, by and large, vegetarian. Hence, there is a wide scope of vegetable farming. And majority of respondents do it but mostly on subsistence basis. Only 37 respondents or 3.58 per cent of the total respondents have vegetable farming as their first occupation. Kori stands first among all the castes constituting about one-fourth of total respondents under this category. About 50 per cent Koris have vegetable farming as their first occupation. Pasi is an other caste with 19 per cent of its respondents having vegetable farming as their first occupation, although they account 35 per cent of total respondents under this category. Brahmins constitute 22 per cent of respondents under vegetable farming, although only 3.4 per cent Brahmin respondents fall under this category. Ahir, Rajput and Kachhi are other castes among which some respondents have vegetable farming as their first and main occupation. Kachhi is a vegetable growing and selling caste in eastern and central Uttar Pradesh. It is their main occupation. They even purchase and resell it in the villages and weekly markets.

Piggeries

Among the non-vegetarian type of occupations, piggeries stands first occupying 7 per cent of total respondents. Piggeries is popular only among scheduled caste. No other caste except one Pasi respondent has piggeries as its first occupation. About 99 per cent of total respondents having piggeries as their first occupation are scheduled castes, although only 31 per cent of scheduled caste respondents are under this category. The reasons for such a discriminate adoption of this occupation is definitely social and cultural. It is an inferior occupation being adopted by low caste people.

Poultry Farming

Only 6 respondents or 0.59 per cent of total respondents have poultry farming as their first occupation. All the 6 respondents are muslims, accounting about 7 per cent of total Shaikh respondents.

Fishing

Like poultry farming, fishing is also the main occupation of only 5 respondents or 0.48 per cent of total respondents but is the main occupation among several castes. About 80 per cent share goes to scheduled caste and Lodh, while 20 per cent goes to others. But in terms of respondents, only 2 scheduled caste, 2 Lodhs have fishing as their first occupation. Here again religious, social and cultural factors are basic determinants.

Service, Labour and other Occupations

Although these are not primary occupations, there are about 20 per cent respondents having these occupations as their main source of income. Similarly there is a reasonable percentage of respondents who have these occupations as their second and third occupation. Moreover, these are the supporting occupations to primary occupations and, therefore, they are also included here. About 13 per cent of total respondents are engaged in labour. The lion's share i.e. about 54 per cent goes to scheduled caste, although only 30 per cent of scheduled caste respondents belong to this category. Respondents with labour as main occupation are found among all the castes, except Rajput, Gujar, Gadariya and Banjara. Pasis, with 22 per cent of their total respondents, constitute about 12 per cent of total respondents under this category, while Brahmin, Shaikh and Ahir respondents account about 6 per cent each to total respondents under this category. Brahmin and Rajput are the two castes accounting together about 64 per cent of the respondents under this category, although only 9.8 per cent and 13.00 per cent Brahmin and Rajput respondents respectively have service as their first occupation. Scheduled Caste is the other caste accounting 17.7 per cent of respondents under this category, although only 4 per cent scheduled caste respondents are in it. Shaikh, Kurmi and Ahir are other castes in service. The nature of service differs from caste to caste. By and large, the high caste respondents are in white collar jobs, while the low caste respondents are in fourth class services or in ministerial jobs.

Castes and their Second Occupations

Crop Farming

Besides their basic occupations, the respondents have their second and third occupations as well in order to support their livelihood, and also to raise their socio-economic status. About 80 per cent of the selected respondents have their second occupation; while the remaining have one and the only occupation (Table 8.3). Here again, crop farming stands first engaging about 27 per cent respondents under this category. But the percentage strength of different castes in this category has got changed considerably. Here Ahir respondents constitute the highest percentage of respondents amounting to 31.5 per cent, although about 54 per cent Ahir respondents belong to this group. Scheduled Caste rank second contributing about 24 per cent of total respondents under this group; although 27 per cent Scheduled Caste respondents have crop farming as their second occupation (Fig. 8.4). The highest percentage of Scheduled Castes under this category is due to the government policy of distribution of land among landless as well as due to leasing out of land by land owners of higher size of holding due to shortage of labourer. Brahmins come third constituting about 12 per cent of respondents falling in this group which shows that the Brahmins who were having their first occupation other than crop farming, have, by and large, crop farming as their second occupation. Rajputs, with 22 per cent of their respondents, constitute 9 per cent of total respondents under this group (Table 8.4). About 15 per cent Jat respondents are under this group, although they account only 1.83 per cent of total respondents under this category. Pasi is a farming caste which accounts only 7.3 per cent of the respondents under this group although more than 30 per cent Pasi respondents have crop farming as their second occupation. Shaikh with 16 per cent of their total respondents, constitute 5.5 per cent of the total respondents in this category. Kurmi, the most efficient cultivator of eastern Uttar Pradesh, have more than one-fifth of its respondents under this category, although they account only 2.74 per cent of respondents under this category.

TABLE 8.3: **Castes and their second occupations (number of respondents)**

Major Castes	Crop Farming	Dairy Farming	Fishing	Vegetable Farming	Poultry Farming	Piggeries	Service	Labour	Others
Brahmin	37	39	—	35	—	—	44	4	—
Rajput	20	23	2	23	—	—	21	3	—
S.C./S.T.	52	10	23	12	4	56	7	29	—
Jat	4	14	—	2	—	—	4	1	1
Nomads	1	9	1	—	—	1	—	3	1
Shaikh	12	23	3	8	19	—	5	2	3
Kurmi	6	9	—	8	—	—	1	2	2
Ahir	69	40	—	14	1	—	1	6	—
Pasi	16	2	4	26	1	—	1	2	—
Kachhi	—	1	—	2	—	—	—	—	—
Lodh	2	2	5	2	1	—	2	3	—
Kori	7	2	—	3	—	—	1	1	—
Other Castes	3	7	—	1	—	—	—	1	5
Total	219	181	38	136	26	57	87	57	12

Dairy Farming

Dairy farming, with 181 respondents or 22.26 per cent of the respondents under this category is next in importance, although it was third in case of first occupations. Ahirs again have come as first with 30.5 per cent Ahir respondents under this category. Brahmins, with more than one-fourth of Brahmin respondents, follow Ahir, constituting about 22 per cent of total respondents under this category. Dairy farming is also very popular among Jats. About 54 per cent Jat respondents, constituting 7.73 per cent of total respondents under this category, have dairy farming as their second occupation. Rajputs constitute more than 12 per cent of total respondents under this category, although 25 per cent Rajput respondents fall under this category. Gujar, Gadariya and Banjara are, by and large, nomadic

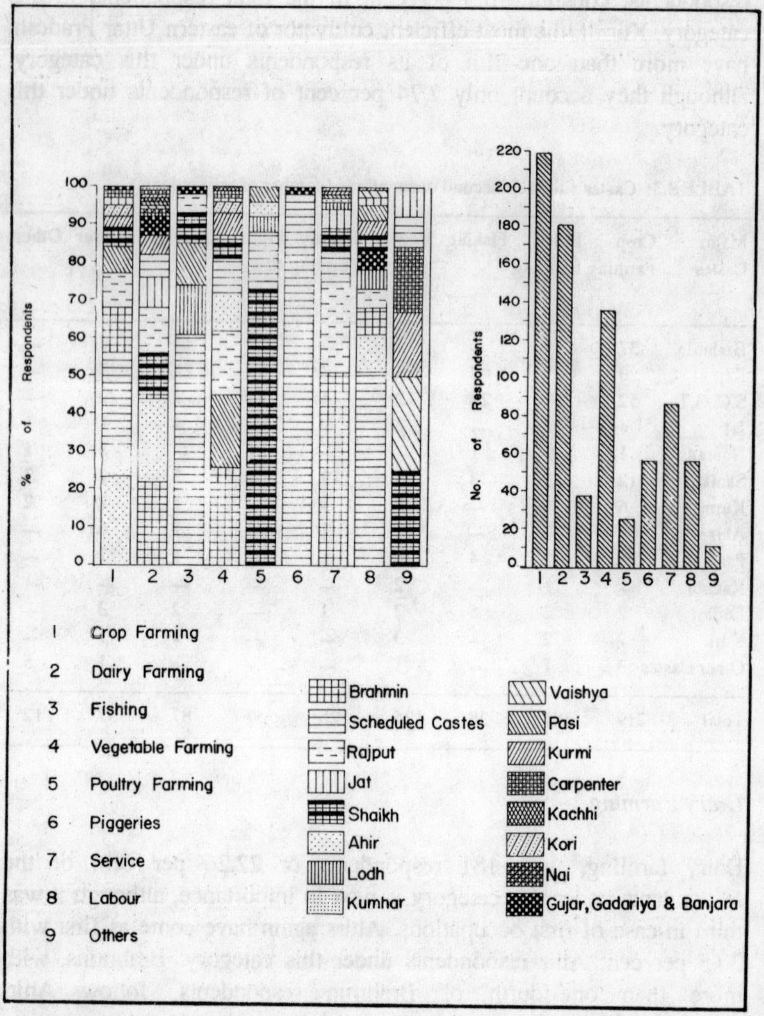

FIG 8.4: Major occupations (II occupation).

castes and they move along with their animals. About 60 per cent respondents of these castes have dairy farming as their second occupation, although they account only 5 per cent of total respondents under this category. Dairy farming is also popular among Shaikhs. About 31 per cent of Shaikh respondents has dairy farming as second occupation, and they account 12.7 per cent of total respondents under this category. Scheduled Caste respondents account about 5.5 per cent of respondents under this category. Kurmis, with 32 per cent of their total respondents, constitute about 5 per cent of the total respondents under this category. Besides these, Kumhar, Carpenter, Pasi and Lodh have also dairy farming as their second occupation. This analysis shows that dairy farming is still popular among traditional dairy farming castes like Ahir, Gujar, Banjara, Jat, etc.

TABLE 8.4: Castes and their third occupations (number of respondents)

Major Castes	Crop Farming	Dairy Farming	Fishing	Vegetable Farming	Poultry Farming	Piggeries	Service	Labour	Others
Brahmin	4	22	4	27	—	—	13	1	8
Rajput	1	17	3	12	1	—	10	—	—
S.C./S.T.	24	5	62	5	8	9	10	1	1
Jat	—	—	—	5	—	—	1	—	1
Nomads	—	1	1	—	1	—	—	—	—
Shaikh	5	8	8	6	25	—	—	—	—
Kurmi	2	4	1	9	—	—	—	1	—
Ahir	10	7	1	18	—	—	19	5	—
Pasi	3	3	4	1	1	1	2	—	—
Kachhi	—	2	—	—	—	—	—	—	—
Lodh	2	—	3	—	—	—	1	—	—
Kori	2	6	—	—	—	—	1	1	—
Other Castes	—	6	1	3	—	—	2	—	—
Total	53	81	88	86	36	10	59	9	10

Vegetable Farming

Vegetables constitute very important component of diet among rural Indians. There are several seasonal vegetables which are grown in the kitchen garden or on their thatched roofs for which there is no data. But personal observations indicate that during rainy season, a lot of vegetables are grown and some of them even continue till winter season. However, due to methodological and procedural problems, the data of such vegetables could not be collected and the present analysis is based on vegetables grown in the field. Table 8.3 show that 136 respondents or more than 22 per cent of the total under this category, have vegetable farming as their second occupation. In this regard Brahmins with less than one-fourth of their total respondents account 25.74 per cent of respondents under this category. Pasi, Kachhis and Kori are other vegetable growing castes, although all the three put together account only 14 per cent of the respondents under this category, but 50, 66.67, 50.0, and 21.3 per cent respondents of the respective castes are under this group. However, in terms of percentage to total respondents under this category, Rajputs follow the Brahmin with 25 per cent of their respondents accounting to about 17 per cent in this category. Scheduled Caste respondents constitute 8.8 per cent respondents of this group, while Shaikhs account 6 per cent with 10 per cent of their total respondents. Ahirs are equally important with 10 per cent of their respondent in this group. Vegetable farming is also popular among Kurmi vegetarian caste with 29 per cent of their respondents having it as their second occupation. Carpenter and Lodh are other castes accounting about 2 per cent of total respondents who have vegetable farming as their second occupation.

Piggeries, popularly known as an occupation of the scheduled caste or low caste people, constitutes about 7.0 per cent of total respondents under this group. Because of religious, social and cultural taboos, it is totally absent among all the castes except scheduled caste and a single respondent among Gadariya. Scheduled caste constitute 98.3 per cent of total respondents under this category, although only 29 per cent scheduled caste respondents have it as their second occupation.

Fishing

Their are 38 respondents or 4.67 per cent of total respondents under this category who have fishing as their second occupation. There are only 6 castes which practice fishing. In terms of percentage contribution to total respondents under this category, scheduled caste rank first accounting 60.5 per cent, although only 12 per cent scheduled caste respondents have fishing as their second occupation. Lodh is a second caste in importance constituting 13 per cent of total respondents under this category, but about 30 per cent Lodhs have fishing as their second occupation. Fishing is popular among Pasis of some region and they constitute about 10 per cent of total respondents under this category and about 8 per cent Pasis have fishing as their second occupation. It is also popular among Shaikhs. Four per cent Shaikhs have fishing as their second occupation accounting to about 8 per cent of total respondents in this category. Some Rajputs also have fishing as their second occupation.

Poultry Farming

By and large, poultry farming is done as household occupation mainly by the women folk. But recently due to larger number of population using egg, and other poultry products; it has gained popularity, and is being practiced on commercial farm basis. In the rural areas it is still the occupation of low caste muslims, although scheduled caste, Pasi and some other Hindu castes have also started it. In urban and sub-urban areas, it has now become the occupation of non-muslims. However, in the present study 19 Shaikhs, accounting to 73 per cent of total respondents under this category, have poultry farming as their second occupation and they are about 25 per cent of Shaikh respondents. Scheduled Caste, constituting 15 per cent of total respondents under this category, stands second; although only 2 per cent of scheduled castes have poultry farming as their second occupation. Besides these, one each among Ahir, Pasi and Lodh respondents has poultry as his second occupation. Herein again we can see the impact of religion and culture on occupation.

Service, Labour and Others

About one-fifth of the total respondents have their second occupation other than farmings. Out of this, about 11 per cent respondents are in service of various kinds. Brahmins and Rajputs take the lead accounting 75 per cent (51 and 24 per cent respectively) respondents under this category, although only 52 per cent respondents of these castes have service as their second occupation. Scheduled Castes and Jats follow the line accounting 8 and 5 per cent of total respondents under this category, although 15 per cent Jat respondents have service as their second occupation. About 7 per cent Shaikh respondents have service as their second occupation. Besides these, some Kurmi, Ahir, Pasi, Lodh and Kori respondents have service as their second occupation constituting together about 9 per cent of total respondents.

Scheduled Caste — the popular labour caste, accounts 50 per cent of respondents under this category. With 15 per cent of its respondents under it, Ahir also account 10 per cent of respondent under this category. Poor Brahmins and Rajputs together account 12 per cent of respondents under this category. In the service, the nature of service varies with castes viz. high caste respondents have better service or white collar job, while low caste people have inferior type of services.

Castes and their Third Occupations

Crop Farming

As the order of occupation in priority decreases, the number of respondents also decreases. Now there are only 432 respondents or about 42 per cent of total respondent under this category but they are found among all the occupations discussed above. Among third occupations, fishing — a non-vegetarian occupation, ranks first accounting more than one-fifth of total respondents under this category (Fig. 8.5). However, crop farming is also important occupation in this group. Fifty three respondents or 12.3 per cent of total respondents are under this category. In crop farming as third occupation scheduled caste rank first constituting 45 per cent of the total respondents under this group. The reason is that besides piggeries and labour as first and second occupation, they do little bit of crop farming on the land given by government or taken on lease basis. About one-fifth of scheduled caste respondents are in it. This accounts 19 per cent of respondents under this category with 17 per cent of their respondents having crop farming as their third occupation. Shaikh with 10 per cent of their respondents account equal percentage to the total respondents under this category. Brahmins come next with only 5 per cent of total respondents under this category. Lodh with 33 per cent, and Pasi and Kori with 20 per cent each together account about 14 per cent of the total respondents under this category.

Dairy Farming

Dairy farming is popular as third occupation also, constituting about 19 per cent of respondents under this category. Brahmins with about 27 per cent of their respondents, constitute almost equal percentage to total respondents in this category. Rajputs with 39 per cent of their respondents and accounting 21 per cent of its respondents falling under

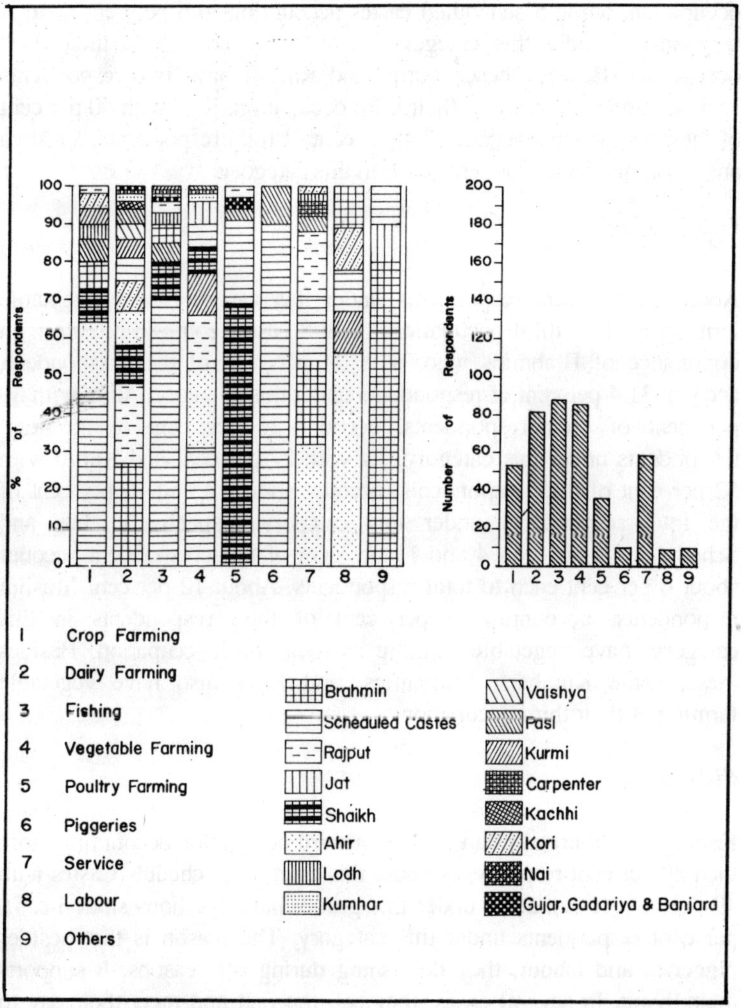

FIG 8.5: Major occupations (III occupation).

it, account almost equal per cent to total respondents under this category. There are 7 Ahirs with dairy farming as their third occupation, while 5 scheduled castes accounting to 6 per cent of total respondents under this category, have dairy farming as their third occupation. Besides these, Kurmi and Kachhi have two respondents each with dairy farming as their third occupation. Kori with 60 per cent of their respondents account 7.4 per cent of total respondents. Vaishya and Lodh have 3 respondents each in this category.

Vegetable Farming

About 20 per cent respondents under this category have vegetable farming as their third occupation. Only 8 castes come in picture with dominance of Brahmins who, with 34 per cent of their respondent, account 31.4 per cent of respondents under this category. Ahirs with 30 per cent of their respondents, account 21 per cent of the total respondents under this category. Jat with 27 per cent and Kurmi with 52 per cent of their respondents, account about 14 and 10 per cent of the total respondents under this category respectively. Jats and Scheduled Castes with 4 and 11 per cent of their respondent account about 6 per cent each to total respondents. About 12 per cent Muslim respondents, accounting 7 per cent of total respondents in this category, have vegetable farming as their third occupation. Besides these, some Kumhars, Carpenters, and Pasis also have vegetable farming as their third occupation.

Fishing

Fishing has emerged as an important third occupation accounting more than 20 per cent respondents under this category. Schedule Castes with 49 per cent respondents under this group, have the lions share i.e. 70 per cent respondents under this category. The reason is that besides piggeries and labour, they do fishing during off seasons. It supports their living. In several cases, they sell the fish and meet their day to day household expenses. Muslims with 15 per cent of their respondents constitute 9 per cent of total respondents. Interestingly some Brahmins and Rajputs also have fishing as their third occupation, although their percentage is insignificant. Pasis with more than one-fourth of Pasi respondents have fishing as their third occupation. The other castes with fishing as their third occupation are Gujar, Gadariya, Banjara,

Kurmi and Ahir, which together account not more than 4 per cent of total respondents under this category. Herein again, the hierarchy of castes and their occupation is reflected.

Poultry Farming

Poultry farming with 8.3 per cent of total respondents in this category, is next important non-vegetarian type of occupation. Only 5 castes have it as their third occupation, but the lion's share goes to Muslims who, with 48 per cent of their respondents, constitute 70 per cent of total respondents under this category. Scheduled Castes with only 6 per cent of their respondents account 22 per cent of total respondents. It shows that poultry farming, as third occupation, is popular among low caste muslims and scheduled castes. The other three castes with one respondent each having poultry farming as third occupation are Rajput, Pasi and Gujar.

Piggeries

It is not very common as third occupation. Only 10 respondents or 2.3 per cent of the total have it as their third occupation. Out of these 9 are scheduled castes and only one belongs to Pasi caste.

Service, Labour and Others

There are 78 respondents or 18 per cent of total respondents with other than primary occupations as their third occupation. The substantial percentage (14%) is among service. In this case Ahirs, constituting 32 per cent of total respondents, come first followed by Brahmins with 22 per cent in service. Scheduled castes with 8 per cent of their respondents rank third accounting 17 per cent of total respondents under this category. Rajputs with 23 per cent of their respondents come next in line. The other castes, having services as their third occupation, are Jat, Lodh and Kori. Only 2 per cent respondents have labour as their third occupation. Only 5 castes emerge into picture out of which Ahir has 5 respondents with labour as third occupation. The remaining four castes have one respondent each with labour as their third occupation.

The following points clearly emerge out of the above discussion of castes and their primary occupations. And they prove our several hypotheses put forth in this study.

(i) Crop farming is the main occupation in the state. The similar pattern emerges in the analysis of caste and primary occupations. Nearly 57.3, 21.2, and 5.1 per cent respondents have crop farming as first, second and third occupation respectively.

(ii) Dairy Farming is next important occupation engaging 380 respondents with their either first, second or third occupation. Out of this nearly 27.6 per cent have dairy farming as their first occupation, while 47.6 and 21.3 per cent have it as their second and third occupation respectively.

(iii) Vegetable farming ranks third in terms of number of respondents engaged. There are 259 respondents accounting 25.07 per cent of total respondents, who have vegetable farming as first, second or third occupation. Nearly 14.29 per cent have dairy farming as their first occupation, while 52.51 and 33.20 per cent have it as their second and third occupation respectively.

(iv) Crop farming is main occupation of high caste respondents. They grow fine grains and more remunerative crops as they produce them for selling purposes. On the contrary, low caste respondents grow coarse grains and produce it on subsistence basis. Similarly there is a considerable difference in agricultural production also. The agricultural productivity, by and large, is higher among high caste respondents as compared to low caste respondents.

(v) In spite of technological development and modernisation of society and agriculture both, there is still distinction and hierarchy of occupation according to caste to a very large extent.

(vi) Dairy farming, by and large, is main occupation of Ahirs, although the other castes also practice it but mainly on subsistence basis. Ahirs practice dairy farming mainly on commercial basis. Nearly one-third of total respondents engaged in dairy farming as their first, second, or third occupation are Ahirs. Among the Ahir respondents nearly 88 per cent are engaged in diary farming as their first, second or third occupation. Out of this, about 54 per cent have it as their first occupation; while about 29 and 5 per cent have it as second or third occupation respectively. Even among Ahir respondents with dairy farming nearly 61 per cent have it as first occupation and 33 and 6 per cent as second and third occupation respectively. Besides occupational hierarchy, there is variation in dairy products also. The production per unit of animals or per head, is higher among Ahir respondents as compared to the respondents of other castes who practice dairy farming. Ahir respondents know the art of keeping dairy cattle, whereas the respondents of other castes lack it. He is also almost a semi-doctor for various diseases of dairy animals.

(vii) Vegetable farming has got considerably changed specially around urban centres. People of different castes have started growing it. But still in rural areas, specially in remote rural areas, it is more popular among Kachhi, Kori, Pasi and Kurmi. Nearly 28 per cent of total respondents engaged in vegetable farming as their first, second and third occupation, belong to these castes. About 48 per cent respondents of these castes have vegetable farming as their first, second and third occupation. Out of this, nearly 32 per cent have dairy farming as their first occupation, whereas about 54 and 14 per cent have it as their second and third occupation respectively. The respondents of these castes know the minute details of vegetable farming. And; therefore; the quality and production of vegetables in the fields of these respondents is better and more as compared to the respondents of other castes.

(viii) By and large, poultry farming is practiced by Shaikhs and that too, by low caste people. The other castes who practice poultry farming belong to lower castes. Nearly 59 per cent Shaikh respondents have poultry farming as their first; second or third occupation, and they constitute nearly 74 per cent of the respondents practicing poultry farming. However, among the poultry farming Shaikh respondents; 12 per cent have it as first occupation; while 38 and 55 per cent practice it as their second or third occupation respectively.

(ix) Piggeries is main occupation of scheduled castes. It is rarely practiced by respondents of other castes. Nearly 58 per cent scheduled castes have piggeries as their first, second or third occupation. They constitute nearly 98 per cent respondents practicing this occupation. Among the piggeries practicing respondents, nearly 53 per cent have it as their first occupation, while 41 and 6 per cent have it as their second and third occupation respectively.

(x) Fishing is common among low caste people whether they belong to Hindu or Muslim religion. Scheduled Castes, Lodh and low caste muslims have it as their first, second and third occupation.

(xi) The poor respondents of higher castes in general and the respondents of low castes in particular, are labourers. They work as agricultural labourer or freelancing labourers. The Scheduled Castes dominate the scene. Nearly 43 per cent Scheduled Castes respondents are labourers having it as their primary, secondary or tertiary source of income. They constitute nearly 51 per cent of total labourer respondents. Among the scheduled caste labourer respondents, 70 per cent have it as their main source of livelihood, while 29 and only one per cent respondents have it as their secondary and tertiary source of income respectively.

(xii) Substantial percentage of respondents in service belong to high castes specially Brahmin and Rajput. A minor percentage of lower caste respondents are in service. A further analysis of respondents in service indicates that higher caste respondents are in superior and better jobs (white collar jobs), while lower caste respondent are in inferior and ministerial services.

REFERENCES

Alexander, J. W., 1963 : *Economic Geography,* New Jersey.

Barth, F. : The 1960 : The system of Social Stratification in Sawat, North Pakistan in E.R. Leach (Ed.) *Aspects of Caste in South India, Ceylone, and N.W. Pakistan.*

Buchanan, R. G., 1959 : Some Reflections on Agricultural Geography, *Annals of the Association of American Geographers,* Vol. 44.

Bailey, F. G., 1963 : Closed Social Stratification in India, *European Journal of Sociology,* No. 1, Vol. XIV.

Berreman, G. D., 1967 : Stratifications, Pluralism and Interaction : A Comparative Analysis of Caste, in Reuch, A.D. and Knight, J. (Eds.) *Caste and Race,* London.

Beteille, A., 1965 : *Caste, Class and Power,* California.

———, 1961 : Closed and Open Stratification, *European Journal of Sociology,* Vol. VII.

———, 1969 : The Politics of Non. Antagonistic Strata, *Contribution to Indian Sociology.*

———, 1974 : *Studies in Agrarian Social Structure,* OUP, London.

Bhat A., 1975 : *Caste, Class and Politics,* Delhi.

Berry, B. J. L., et. al. 1976 : *The Geography of Economic Systems,* New Jersey.

Crooke, W., 1974 : *The Tribes and Castes of the Northern India* Vol. I-IV.

Caplo, T., 1954 : *The Sociology of Work*, Illinois.

Dutta, N. K., 1931 : *Origin and Growth of Caste in India*, Calcutta.

Dore, R. P., 1958 : *City Life in Japan : A Study of Tokyo Ward*, Berkely.

Ghurye, G. S., 1961 : *Caste, Class and Occupation*, Bombay.

Grigg, D. B., 1974 : *The Agricultural Systems of the World : An Evolutionary Approach*, London.

———, 1984 : *An Introduction to Agricultural Geography*, London.

Gadgil, D. R. 1945 : *Poona : A Socio-Economic Survey*, Part I & II.

Hutton, J. S. 1946 : *Caste in India*, Oxford.

Jones, C. E. and Darkenwald, G. G., 1962 : *Economic Geography*, New York.

Kahi, J. A., 1957 : *The American Caste Structure*

Kosambi, D. D. and Sharma, R.S. (Eds.) 1958 : *Indian Society : Historical Probings in Memory of D.D. Kosambi*, New Delhi.

Kumar, A., 1985 : *Impact of Caste on Primary Occupation* (Unpublished M.A. Dissertation, Delhi University).

Leach, E. R. (Ed.) 1960 : *Aspects of Caste to South India, Ceylone and North West Pakistan*, Cambridge.

Louis Dumont, *Home-Hierarchicus : The Caste System and Its Implications* (Trans. Marx Samisbury).

Mohammad, N., 1978 : *Agricultural Landuse in India*, Delhi.

———, 1981 : *Perspectives in Agricultural Geography*, Volume I to V, Delhi.

———, 1981 : Socio-economic Determinants of Diffusion of Agricultural Innovations, in N. Mohammad, (Ed.) *Perspectives in Agricultural Geography*, Vol. III.

Mukherjee, A. B., 1980 : *The Chamars of Uttar Pradesh*, Delhi.

Nasfield, J. C., *Caste System of Northwestern Provinces and Oudh.*

Nijhawan, P., 1986 : *Impact of Caste on Primary Occupation* (Unpublished M.A. Dissertation, Delhi University)

Omvedt, G. (Ed.) 1982 : *Land, Caste and Politics in Indian States*, Delhi.

Pandit, M., 1979 : Caste and Class in Maharashtra, *Economic and Political Weekly*, Annual No. XIV, No. 780.

Risley, H. H., 1969 : *The People of India*, 2nd Ed. Delhi.

Russel, R. V. and Lal, H., 1975 : *The Tribes and Castes of Central Provinces of India*, Vol. I-IV (Reprint).

Schwartzberg, J. E., 1965 : The Distribution of Selected Caste in the North Indian Plain, *Geographical Review*, Vol. 45, No. 4.

———, 1968 : Caste Regions of the Northern Indian Plains, in Singer, M. and Bernard, S. H. (Eds.) *Structure and Change in Indian Society*, Chicago.

———, 1978 : *A Historical Atlas of South East Asia*, London.

Sinhas, S., 1982 : *Caste System : Myths, Reality Challenge*, N. Delhi.

Sharma, R. S., 1958 : *Sudras in Ancient India*, Delhi.

Sharma, K. K., 1984 : *Impact of Caste on Primary Occupations* (Unpublished M.A. Dissertation, Delhi University).

Sharma, D. D., 1983 : *Impact of Caste on Primary Occupations* (Unpublished M.A. Dissertation, Delhi University).

Singh, Y., 1968 : Caste and Class : Some Aspects of Continuity and Change, *Sociological Bulletin*, No. 2, Vol. XVII.

Srinivas, M. N., 1966 : *Social Change in Modern India*, Berkeley.

Singh, J., 1970 : *Agricultural Atlas of India : A Geographical Analysis*, Kurukshetra.

Singh, R., 1974 : Agrarian Social Structure and Peasant Unrest : A Study in Land Grab Movement in Basti, East U.P. *Sociological Bulletin*, No. 23.

Sharma, K. L., 1972 : Levels of Mobility in Caste Structure : A Hypothetical Refinement, *The Indian Journal of Social Work*, Vol. XXXII.

Thoman, R. S. et.al., 1968 : *The Geography of Economic Activity*, N. York.

9

AGRICULTURAL CAPITALISATION AS A SPATIAL COMPLEX

HOWARD F. GREGOR

AGRICULTURAL GEOGRAPHERS have traditionally emphasised the visual aspects of the rural landscape, but in so doing could well be ignoring one of the most important spatial elements of agriculture today: capitalisation. The application of capital to the farming process is now well recognised as a growing worldwide phenomenon. In the industrialised countries it has brought huge production gains but also created social and resource problems whose full seriousness is only beginning to be realised; in the more agrarian countries it is regarded as absolutely essential, although there is a debate over the form and depth of capitalisation to be sought. Yet, despite this prominence, little attention has been given to agricultural capitalisation in a spatial way, and that has been presented mostly in general world regional terms (Herlemann, 1954). There seems no reason, therefore, why a more detailed and localised spatial analysis cannot be made of the process, particularly where capitalisation is well advanced and sufficient data are available. In these respects, the United States, and in particular its south-western corner where some of the highest as well as the lowest levels of capitalisation in the country are attained, provides an excellent study area.

Research Strategy

Emphasis on the process of agricultural capitalisation rather than its end products puts special emphasis on the need for knowing its structure. Factor analysis provides one way of deciphering it, as based on the nature of the various variable clusters (factors), the comparative weights of the variables within each factor, and the relative importance of the factors to each other. Higher-order factor analysis, in turn, searches for still more and broader structures, with the factors previously obtained now becoming the variables.[1] No less important to the geographer is the ability of the factoring procedure to assign weights (factor scores) to the data-reporting units so that the areal distribution of each factor can also be determined. Thus the complex under study acquires a spatial character, which also aids in further understanding the nature of the functional dimension.

Yet the factoring procedure, like any statistical technique, is a neutral element and for best results must be applied to variables that are pertinent, comprehensive, and so designed that all known aspects of the complex being studied are, as much as possible, covered proportionate to their importance. Thirty variables were selected. They were computed from country data provided by the federal agricultural census for the States of Arizona, California, Hawaii, Nevada and Utah. (U.S. Bureau, 1972). The majority of the variables were obtained by first computing the dollar amounts of capital expended and invested in each of the principal forms of farm capital, here understood in its more common and extended meaning: land, labour, machinery and equipment, crops and livestock. Each of these inputs was then related individually to the supply of farmland, amount of labour, total agricultural expenditures and total agricultural investments. Large capital inputs put a heavy premium on good management, and to represent this the "net value of farm products sold" was calculated and also applied to the four denominators. An agricultural specialisation variable, assumed to be another pertinent indicator of management efficiency, was also included. It was based on the proportion of farms receiving half or more of their sales revenue from a single product or group of products. Also included in the variable mix was one that takes into account the intensity of capital application by the operational unit, the farm. This variable was represented by the percentage of farms with less than 260 acres (100 hectares), the reasoning being that smaller units can generally be expected to increase their capital intensity to compensate for smaller acreages.

Capitalisation Patterns (First-Order)

Six major factors were extracted from the Spearman correlation matrix of the variables.[2] The most important, and accounting for almost a quarter of the distributional variance of all of the variables, is one of heavy concentration on cropping. It evidences the superiority of crops over livestock in the capitalisation process, for cropping is particularly receptive to increased non-farm, or industrial, inputs. So important is this cropping dimension that all the indicators of high cropping levels in the capitalisation model are represented: intensity of crop investments and expenditures, proportion of available capital devoted to crops, and ratio of crop investments and expenditures to agricultural workers (Table 9.1). Conversely, a very low rating is given to two livestock capitalisation indicators, percentage of investments in livestock and livestock investments per worker.

Still not all of the cropping variables are of equal significance. By using an oblique instead of the traditional orthogonal rotation of the factors in the factoring process, one can determine the amount of correlation between factors and in what variables it exists. Thus it is possible for some variables to be more directly related to the given factor, whereas others gain their importance in correlations through other correlated factors. These differences are shown in a comparison of the weights (loadings) of the variables on Factor I under the "Factor Pattern", and "Factor Structure" titles in Table 9.1. The factor pattern shows best what variables are highly involved in what clusters and the degree of involvement of each variable in the clusters, as expressed in the loadings of the variables on each factor. When squared, these loadings give the direct contribution of the factor to the variance of the variable. It is apparent that on this basis the cropping factor is one more of scale and orientation than of intensity, for the two crop-intensity variables have only half the value of the two leading variables.

The differences among the cropping variables are much smaller in the factor structure, which, although not displaying the factor pattern as well, does give the exact correlations of the variables with the factors. Also, when squared, the structure coefficient indicates the total contribution a factor makes to the variance of a variable, both directly and indirectly through the other factors. When the squared structure and pattern coefficients are compared, it can be seen that the cropping intensity variables (crop investments per acre; crop expenditures per acre) are the ones that most improve their factor structure position through the indirect contribution. This relationship can also be easily noted in the high correlations of the two variables with a factor that emphasises landuse intensity, as shown by the spillover into the Factor II column of the factor structure.

TABLE 9.1: First-Order factors for agricultural capitalisation*

Variable	Factor Pattern						Factor Structure					
	I	II	III	IV	V	VI	I	II	III	IV	V	VI
1	2	3	4	5	6	7	8	9	10	11	12	13
Crop investments/worker	.95						0.96					
Crop expenditure/worker	.95						.95					
% expenditures for crops	.88						.94					
% investments in crops	.81						.93	.59				
Crop investments/acre	.65						.89	.78				
Crop expenditures/acre	.61						.87	.81				
% investments in livestock	−.58						−.66					−.56
Livestock investments/worker	−.58						−.80	−.66				
Livestock investments/acre		.91						.85				
Machinery and equip. investments/acre		.77					.71	.91				
% farms specialised		.70					.56	.84				
Labour expenditures/acre		.70					.70	.92				
Livestock expenditures/acre		.69					.52	.85				
Land and bldg. investments/acre		.68					.63	.85				

(Contd.)

TABLE 9.1 *(Contd.)*

1	2	3	4	5	6	7	8	9	10	11	12	13
% farms less than 260 acres		.67					.52	.82				
Machine-hire and fuel expenditures/acre		.65					.79	.87				
Net value products sold/$ expenditures			1.01						.98			
Net value products sold/worker			1.01						.98			
Net value products sold/$ investment			1.00						.99			
Net value products sold/acre			.91						.93			
Livestock expenditures/worker				.96						.95		
% expenditures for livestock				.91						.94		
% expenditures for labour				−.94						−.93		
Machinery and equip. investments/worker					.94						.91	
Machinery and fuel expenditures/worker					.79						.87	
% expenditures for machine hire and fuel					.72						.84	
% investments in machinery and equip.					.60	−.52					.65	−.72
Land and bldg. investments/worker						.79						.77
% investments in land and bldgs.						.78						.90
Labour expenditures/worker						.70						.66
Sums of squares	5.64	5.39	4.04	3.34	2.90	2.48						
Approximate % of variance	23.7	22.7	17.0	14.0	12.2	10.4						

*Based on *Sub-program Factor, Statistical Package for the Social Sciences* (SPSS). Loadings less than the absolute value of 50 are deleted.

The areal distribution of cropping dominance is clearly a peripheral pattern for the Pacific Southwest States, reflecting perhaps more the extensive and hostile mountain and desert landscapes of the interior than the more westerly and southerly valleys in California and Arizona, where irrigation water is more available and growing seasons are more favourable (Fig. 9.1).[3] As an American State, Hawaii also belongs to this favoured rim.

The relatively more massive than intensive character of the cropping factor shows itself in the variations within the more favoured areas. The peak in the northern Sacramento Valley can be attributed to the almost monocultural emphasis on rice and the extensive mechanisation of its production system. Here is one of the three most important rice-producing areas in the United States, favoured by a hot-summer climate and a heavy soil with underlying hardpan that retains surface water for the crop but also discourages the raising of others. The ratio of cropping investments and expenditures to workers is also increased on the labour side by the heavy investments in machinery, the value on a typical rice farm often exceeding the investment in land (Gregor, 1974). In the same category is the centre of high cropping capitalisation in Hawaii, particularly on the island of Hawaii. Heavy concentration on the raising of sugarcane has created highly productive and extensive monocultural landscapes, at the same time that labour supplies have drastically diminished. Sugarcane raising here is now the most mechanised in the world.

In the remaining cropping centres of note, the oases in the warm deserts of south-eastern California and southern Arizona and the coastal area just south of San Francisco Bay, the high ratio of cropping capitalisation to agricultural workers is especially due to the amount of capitalisation, for the crops raised there are numerous and labour-demanding. The supplying of eastern U.S. markets with vegetables and fruits has long been a primary function of the southern California and southern Arizona areas, and the twin stimuli of nearby and more distant urban markets have made both market gardening and truck farming highly profitable south of the Bay Area.

Intensive landuse easily characterises the second most important capitalisation factor. With cropping dominance, it helps account for almost half of the variance of all of the variables in the capitalisation model and underlines even more the attraction of capitalisation to the best farmland. But the character of this attraction is much more diverse, with livestock, mechanisation and labour being the principal

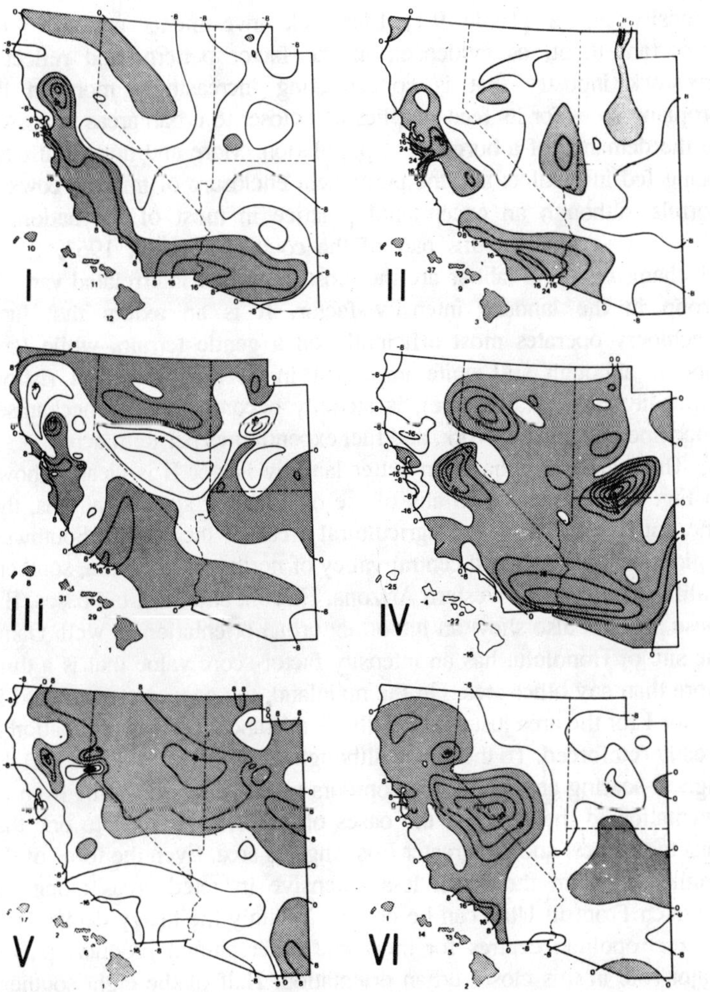

FIG. 9.1: Capitalisation patterns as indicated by positive factor scores (shaded).

9.1-I Cropping dominance	*9.1-II Intensive landuse*
9.1-III Management efficiency	*9.1-IV Livestock dominance*
9.1-V Mechanisation dominance	*9.1-VI Land dominance*

Note: Scores are rounded off to one place beyond the decimal point and the decimal has been deleted. Value gradations are of equal interval. Hawaiian Islands are positioned next to the southern California coast and are of the same scale.

intensity inputs (Table 9.1) Livestock investments are the most important input, as evidenced in the factor pattern, and reflect a livestock industry that is concentrating increasingly more in the cropland areas for its feed supplies and closer to urban areas in answer to the demands of a burgeoning population. More and more cattle are being fed in feedlots and the permanent enclosure of milking cows in corrals, although an exceptional practice in most of the nation, is already traditional in this part of the country (Gregor, 1963, 1974). Mechanisation and labour are the most important interrelated variable group in the landuse intensity factor. It is an axiom that farm machinery operates most efficiently on a gentle terrain, while farm labour, although still quite important in its manual aspect (labour expenditures per farm acre), is steadily becoming more mechanised (machine hire, customwork, and fuel expenditures per farm acre).

The marked emphasis on better land by Factor II is clearly shown in Fig. 9.1-II. Better than any of the other factor spatial patterns, this one outlines the principal agricultural areas of the Pacific Southwest region: central coast and central valley of northern California, southern California and south-western Arizona, Hawaii, and the Utah oases. But these patterns also show an important urban orientation as well. Oahu, the site of Honolulu, has an intensity factor-score value that is a third more than any other area. On the mainland, the concentration noted in Factor I for the area just south of the San Francisco Bay conurbation is greatly reinforced. To the south, although the Imperial valley is still the highest-scoring area there, the contours on the map now shift from an orientation to the east and the oases of southern Arizona to one that runs to the west and the greater Los Angeles area. Even the force of the smaller cities in the much less extensive irrigated areas along the Wasatch Front in Utah can be noted. Certainly the heavy demands of the metropolitan centres for meat and other animal products play a major role in this closer urban orientation. Half of the eight southern California countries (including Imperial Country) sell a greater value of livestock and livestock products than they do of crops. Conversely, the greater distance from large urban centres and the comparatively minor place of livestock farming in the agricultural economy of the northern Sacramento Valley cause the area to appear as a hole in the positive-score surface of Fig. 9.1-II.

It is also impossible to ignore the role of cropping in the urban influence on capitalisation, particularly in much of western United States where restrictions on water supply forced an early spatial association of both agricultural and urban populations. Indeed, a brief consideration of the nature of the variables in Factor II shows that all of them except the livestock intensity variable depend much more on cropping than livestock raising, a fact well supported by the extensive correlation of Factor II variables with Factor I in the factor structure (Table 9.1). The capitalisation factor of intensive landuse may thus be viewed not so much as a factor in its own right as one that was divorced from its "most natural" variables in the factoring process simply because of the overwhelming attraction of capitalisation to cropping. In this sense, Factor II becomes another, albeit indirect, powerful indicator of this affinity.

Although only third in importance among the agricultural capitalisation factors, the one designating management efficiency, Factor III, is perhaps most interesting in that it contradicts the general belief in a very close correlation between efficiency and volume of farm production. All four of the factor variables, which are based on the ratio of the net value of farm products sold to the amount of dollar expenditures, dollar investments, agricultural workers, and farmland, correlated closely with each other but to the exclusion of all other factors of agricultural industrialisation, including Factor II, intensive landuse (Table 9.1). Further, the poorest showing of the four variables representing management efficiency is that of the use of land.

Spatially this discrepancy between efficiency and productivity is shown most dramatically in the negative showing of the most productive area in the entire Pacific south-western region, the San Joaquin valley, and in the superior displays of the most important urban area, where competition with agricultural land is most severe (Fig. 9.1-III). Why some of the most productive land in the nation is at least not as efficiently managed as that nearer cities can be traced to a variety of reasons. Cotton, the principal money crop of the San Joaquin valley at the time for which data were available for this study (1969), was subject to some insect damage, over mechanisation, a common failing of smaller farming operations, and over optimum size, a problem by no means rare for larger farms, are also contributors to lower efficiency levels, for the San Joaquin has large numbers of both large and small farms. But perhaps even more important is the probability that government policies and regulations may inadvertently have also contributed to operational inefficiency. Special farm tax provisions have been designed to stimulate agricultural investment, but they also have allowed corporations and wealthy individual investors to generate artificial losses on their farm investments, which are then used to offset or shelter from taxes their non-farm income. A good deal of overplanting and overproduction, with its negative income effects, may well have resulted, not to mention the depressing effects on the operations of the smaller operators who cannot take advantage of such loopholes. There is also the probability, according to some, that irrigation water has been priced too cheaply by State irrigation districts and the federal Bureau of Reclamation, thereby discouraging a potent means of encouraging more efficient farming practices (Gregor 1979).

But urbanisation also has its stimuli for greater farming efficiency. Honolulu, Los Angeles, and the San Francisco Bay Area all exert tremendous pressures on farming through their inflationary effects on land prices, but they also provide lucrative markets for both livestock and crop products, as already noted in the discussion of Factors I and II. The increased cost of land is in itself a negative stimulant for higher farming efficiency, at least up to the point where even the highest quality of operations can no longer compensate for increased costs. These urban centres are also the prime sources of non-farm inputs, so that distance even in such a highly capitalised agricultural area as this one becomes a major locative force in farming operations. Phoenix and Tucson in Arizona and the Salt Lake City urban complex in Utah also evidence this urban "pull" to the detriment of significant but more outlying agricultural land, even though the aggregate data of the counties, particularly the larger units in Arizona, may still be concealing some of the areal contrast.

Not all major production areas are poorly represented by the management efficiency factor, nor are areas with only moderate agricultural production and no major urban centres completely excluded. The important Sacramento Valley portion of California's Central Valley scores as highly on the factor as the Los Angeles area and thus stands in marked contrast to the negative ranking of the core areas of the San Joaquin Valley. Smaller irrigated areas in the more isolated parts of northern California and Nevada and of southern Utah and northern Arizona equal and in a few cases even exceed the factor scores accumulated in most of central and southern California, as well as southern Arizona. All of this merely bears out again the observation by Thunen that profitability need not necessarily be a direct function of market distance. On the other hand, where isolation combines with rugged terrain and severe climate, as in most of northern California, or joins with greater dryness, as in southern Nevada, their scores on the management efficiency factor fall to some of their lowest levels. The lowest figure, through, in Utah, is not so much a reflection of a stringent physical environment as it is of cultural legacy. When the Mormons, members of the Church of Jesus Christ of Latter Day Saints, settled in Utah, they established villages instead of isolated farms and a land-holding pattern that was dispersed rather than concentrated so as to ensure a variety of land quality and thus encourage diversification. All this was done to promote group solidarity and economic self-sufficiency, and although these measures succeeded admirably, the small and dispersed fields resulting from these efforts are now a major obstacle to the farmer who must sell his products and use his machinery as efficiently as possible in order to survive.

The last three important factors of agricultural industrialisation extracted by the factoring process refer to three types of farm capital already noted, livestock, machinery and equipment, and land; now, however, they are characterised more by their dominance of the capitalisation process than their intensity of capitalisation. This dominance is one of heavy capitalisation per worker and significant expenditure of the total capital resource on the type of farm capital in question (Table 9.1). Livestock raising is the most important of the three factors, with 14 per cent of the variance. It is also the most widely distributed, being the traditional economic adaptation to the large areas in the Pacific Southwest that are only marginal in quality even for livestock grazing (Fig. 9.1-IV). Another indication of this more extensive operational character is the sharp contrast between the scoring patterns of the more metropolitan areas on the intensive type of livestock capitalisation, as represented by Factor II, and those on the more extensive type. For the first type, the more urbanised areas rank among the highest; for the more extensive type, the same areas definitely rank among the lowest. The crop-dominated Sacramento Valley also has a very low rating.

But concentrations of this more operationally extensive, yet also more dominant, kind of livestock activity within its overall area of distribution also show that this type need not necessarily be both extensive and only marginally productive, as is usually assumed. Irrigation has made it possible to expand greatly the feed supply in certain favoured areas, as clearly indicated by the high values in southern Utah, north-western Nevada, east-central California, and in south-eastern California and southern Arizona. In these areas, then, the dominance of livestock-raising has actually been due more to the magnitude of capital expenditures than to the sparsity of population and lack of economic alternatives. In south-eastern California and southern Arizona, this volume of capitalisation is so great that the area ranks high on both intensive and extensive livestock-raising characteristics (Fig. 9.1-II and 9.1-IV).

Factor V, with approximately 12 per cent of the variance, shows the same two-fold distribution aspect for mechanisation as does the factor representing livestock dominance (Fig. 9.1-V). Again, it is largely the areas of sparse population and low agricultural productivity that stand out, but also included are a few places where the level of capitalisation is significantly high. Most extensive is the area extending from the mountains of north-eastern California eastward across most of Nevada and Utah. That this large area representative of mechanisation dominance of the capitalisation process does not also encompass the territory in the south included under Factor IV may probably be attributed more to the greater importance of labour-intensive agriculture in the irrigated oases there than to the drier climate.

Although the emphasis is on the production of less labour-demanding feed and grain crops for the support of cattle on adjacent and extensive rangelands, ranchers have also been steadily increasing their irrigation, with attendant expansion of their inventories of feed-handling equipment. Where irrigation supplies are especially sizable and good pasturage is nearby, investments in and expenditures for machinery and other equipment have expanded even more, as indicated along the eastern front of the Sierra Nevada. The highest levels of this capitalisation are most impressive, however, on the large mechanised rice farms of the Sacramento Valley in California and on the heavily mechanised sugar plantations on the island of Kauai in Hawaii, both areas of heavy and valuable crop production. The not-insignificant production of the oases of central Utah may also be noted as a centre of mechanisation dominance of the capitalisation process, but here it appears to be more a symptom of problems than of success, as can be observed in a comparison with the very low level of managerial efficiency indicated by Factor III for this region. Overmechanisation, and particularly on small farms, is now recognised as a widespread problem in the agriculture of both advanced and developing countries. A final observation may also be made of the low representation of this factor in the areas around the major urban centres: none of the six factors representing agricultural capitalisation so conspicuously avoids these areas in degree and areal extent as this one. Here economic stimulus and good physical environments combine to form the greatest barrier to this type of mechanisation capitalisation.

As agricultural capitalisation has advanced, land has become relatively less important, and this is reflected in the small contribution of Factor VI to the total variance, about 10 per cent. Where land does figure strongly in the agricultural industrialisation process, it more often than not reflects environments that discourage the intensive application of other farming inputs. Much of the mountain land of central and northern California, the deserts of southern Nevada, and the plateaus of northern Arizona belong to this group (Fig. 9.1-VI). The equally good showing of the land factor in highly productive farming areas on the margins of the major metropolitan areas of California and Hawaii is undoubtedly at least a partial result of the distortion of the relationship between land values and agriculture by urban demands, although its stimulating effect on agricultural productivity and efficiency, as already noted, should not be underestimated. Similarly, the popularity of second-home properties has added an important non-agricultural influence on the land values in scenic areas that are not too distant from the larger urban centres, particularly the Sierra Nevada of California.

Capitalisation Patterns (Second-Order)

By this time it is obvious from both factor patterns and their spatial distributions that despite the delineation and ordering of six capitalisation factors, there are still enough interrelationships to form the basis of another and broader capitalisation structure of second-order factors. This can be obtained by repeating the procedure used to obtain the first-order factors, with the first-order factors now becoming the variables. The intercorrelations of Factors I, II, and III are important enough to cause them to form the most important of the second-order capitalisation factors (Tables 9.2 and 9.3). Cropping and intensive landuse are the most important characteristics, although good management scores almost as well as either of those two. The spatial distribution of the factor is one embracing all significant agricultural areas in the Pacific Southwest region, but with the greatest emphasis in the areas closest to the metropolitan centres (Fig. 9.2-I).

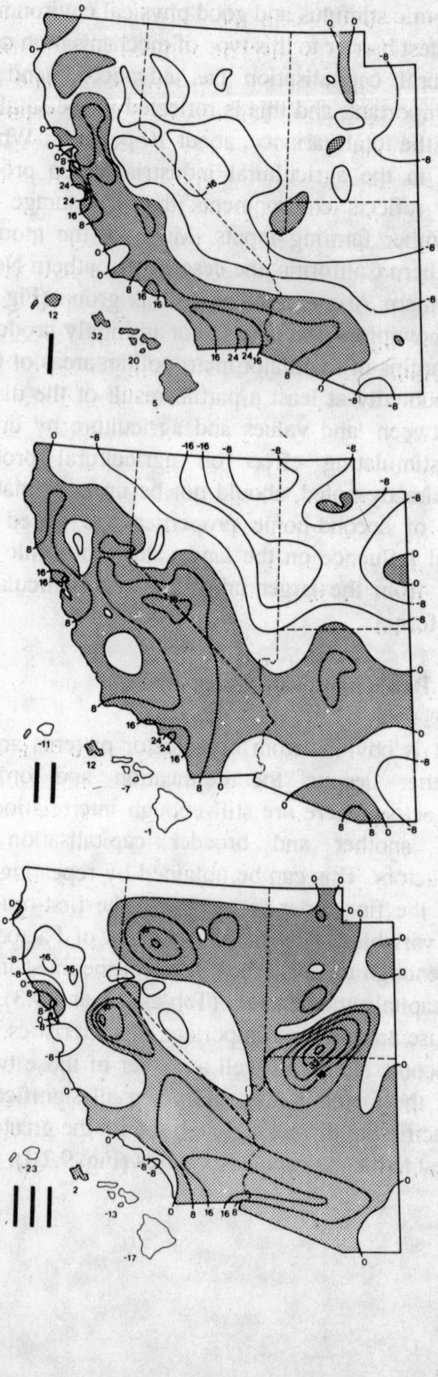

FIG. 9.2: Second-order capitalisation patterns as indicated by positive factor scores (shaded)
9.2-I—*Cropping dominance, intensive landuse, high management efficiency.*
9.2-II—*Land dominance, little mechanisation emphasis.*
9.2-III—*Livestock dominance. Other explanations given in Fig. 9.1.*

TABLE 9.2: First-order factor correlation matrix

Factor	I	II	III	IV	V	VI
I	1.00					
II	.47	1.00				
III	.32	.27	1.00			
IV	−.21	−.08	−.00	1.00		
V	−.09	−.27	−.17	.03	1.00	
VI	−.06	−.08	−.05	−.07	−.30	1.00

TABLE 9.3: Second-order factors for agricultural capitalisation*

	Factor Pattern			Factor Structure		
Variable	I_2	II_2	III_2	I_2	II_2	III_2
Factor II	.78			.79		
Factor I	.75			.75		
Factor III	.69			.68		
Factor VI		.83			.82	
Factor V		−.77			−.79	
Factor IV			.94			.94
Sums of squares	1.79	1.30	1.07			
Approximate % of Variance	43.0	31.3	25.6			

* Based on *Subprogram Factor, Statistical Package for the Social Sciences* (SPSS). Loadings less than the absolute value of .50 were deleted.

Despite its importance, though, Factor I_2 does not dominate the second order complex. Two additional factors result from the collapsing of the original factors, and they total about 57 per cent of the variance. Factor II_2 accounts for 31 per cent of the total variance and is bipolar, contributing strongly to both the positive variance of first-order Factor VI (land dominance) and the negative variance of first-order Factor V (mechanisation dominance). The spatial result of this combination is a pattern that encompasses lands of both good and poor agricultural environments, although much of the most productive farmland (Central Valley of California, southern interior cases) is not represented and the highest scoring areas are again in the vicinities of large urban centres (Fig. 9.2-II). Factor III_2 accounts for approximately a quarter of the variance, and is practically a duplicate of first-order Factor IV, livestock dominance (Table 9.3 and Fig. 9.2-III). Thus, despite being the smallest contributor of the three second-order factors to the total variance, this factor represents the first-order factor that contributes the most to the variance of the second-order capitalisation structure.

Capitalisation Patterns (Third-Order)

Higher-order factoring normally ceases when all correlations between the variables (factors) are eliminated. This stage was reached with the extraction of the three second-order factors of agricultural capitalisation. Still there seems nothing to prevent one, at least from a methodological viewpoint, from factoring the now essentially independent factors so as to arrive at a structure that provides a single-factor summary of agricultural capitalisation.

This was achieved with the extraction of a factor that more than any previous one emphasises the role of urban influence. Factors I_2 and II_2 become the principal variables, whereas Factor III_2, the only one with little urban preference, assumes a negative value (Table 9.4).

TABLE 9.4: **Third-order factor for agricultural capitalisation***

	Factor I_2	Factor II_2	Factor III_2
Factor I_3	.73	.62	–.39

* **Based on *Subprogram Factor, Statistical Package for the Social Sciences* (SPSS).**

In its areal distribution, this summation factor shows a close correlation between urban stimulus and favourable physical environment. Factor scores around the metropolitan areas not only easily surpass the value of other areas, they increase rapidly toward the southerly, and milder, latitudes (Fig. 9.3). But the urban factor still remains primary, if one can judge by the directional trend of the factor-score contours. In central California the highest-scoring areas of the rich Central Valley are not so much in the hearts of the productive Sacramento and San Joaquin valleys as they are somewhere between them (i.e. nearest the Bay Area conurbation). In southern California the contours show a decided gradient from the more urbanised coast to the eastern interior despite the important agricultural production there of southern Arizona. And in Hawaii, the urban-to-rural gradient is even steeper, in spite of the fact that Oahu, with Honolulu, does not contribute to the preponderance of the State's agricultural production.

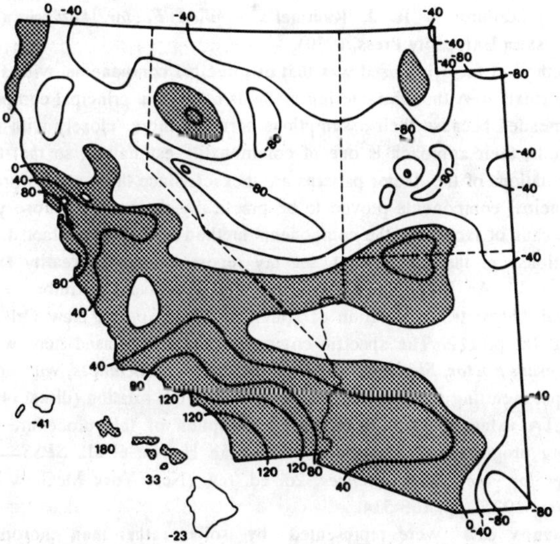

FIG. 9.3: **Third-order capitalisation patterns of agricultural capitalisation, as indicated by positive factor scores (shaded). Scores are rounded off to two places beyond the decimal point and the decimal has been deleted. Other explanations given in Fig. 9.1.**

Conclusion

Few areas in the Pacific Southwest States are without some extensive agricultural capitalisation, but only the better farmland and areas near metropolitan centres are characterised by the most important elements, emphasis on cropping, intensive landuse, and managerial efficiency. Urban proximity, though, is even more an attraction for these components than is farmland, so that the overall distribution of capitalisation is more one of larger cities than principal agricultural areas. It is also a pattern within which urban annexation of farmland is continuing at a rapid rate. Solving these spatial contractions may well be the ultimate test of the capitalisation process.

NOTES

1. **Most literature on factor analysis is quite technical and theoretical. An excellent**

source for the beginner, and for anyone interested primarily in the applied aspects of the procedure, is R. J. Rummel's *Applied Factor Analysis* (Evanston: Northwestern University Press, 1970).

2. The method of factoring used was that of principal components, with a moderate oblique rotation. Although factoring methods other than principal components are recommended because their assumptions correlate more closely with real-world conditions, their approach is one of communality estimation, so that the spatial representations of the factor patterns are inexact. Since the patterns produced by the principal components proved to be practically the same as those yielded by other means of factoring, the components method was finally selected. For other illustrations of the same result, see my report, "Factorial Reality and Spatial Exactness — An Empirical View," in R. C. West and C. Kimber (eds.), *AAG Program Abstracts,* Association of American Geographers, New Orleans, April 9-13, 1978, p. 112. The specific computer programme used here was that of *Subprogram Factor, Statistical Package for the Social Sciences,* with options PAI (principal factoring without iteration) and OBLIQUE rotation (direct oblimin and a DELTA value of zero). A concise description of this procedure and other factoring programmes is provided in Norman H. Nie et al., *SPSS—Statistical Package for the Social Sciences*, 2d. ed. rev. (New York McGraw-Hill Book Company, 1975), pp. 468-514.
3. The country data were represented by isoline rather than choropleth maps because of their better delineation of the important agricultural areas. The principal agricultural centres of the counties were selected as control points. Where centres are few, as they commonly are in the less favourable areas of the interior (especially Nevada, northern Arizona, extreme northern California, and southern and western Utah), the better agricultural districts are exaggerated.

REFERENCES

Gregor H. F., 1974 : "An Agricultural Typology of California", Vol. 4 of *Geography of World Agriculture*, ed. Gyorgy Enyedi; (Budapest : Akademiai Kiado, p. 55.

———, 1963 : "Industrialised Drylot Dairying : An Overview", *Economic Geography*, Vol. 39, pp. 299-318.

———, "The Large Farm as a Stereotype : A Look at the Pacific Southwest", *Economic Geography*, Vol. 55, pp. 78-79.

Herlemann, H. H., 1954 : "Techniesierungsstufen der Landwirtschaft," *Berichte uber Landwritschaft*, Neue Folge, Vol. 32, p. 335.

Rummel, R. J., 1970 : *Applied Factor Analysis*, Evanston : Northwestern University Press.

U. S. Bureau of the Census, 1969 : *Census of Agriculture*, Volume 1. *Area Reports*, Section 2, *County Data.* Washington, D.C. : U.S. Government Printing Office, 1972.

10

SOCIAL CONTEXT OF A MINOR IRRIGATION SYSTEM IN THE ARID ZONES

L. P. BHARARA

THROUGHOUT THE arid tropics, irrigation for agriculture has become a major feature of socio-economic development. The valunerability of India's arid agriculture owing to the vagaries of the monsoon has made it imperative to accord irrigation a high place in the national development plants.[1] Among all the India irrigation systems, development of minor irrigation has made comparatively more substantial progress.[2] Open dugwell-irrigation system makes a large contribution to the irrigated acreage of arid regions.[3] With the implementation of the Drought Prone Areas Programme in the arid zone of Rajasthan, the development and management of the minor irrigation system has received special emphasis.[4]

Symbolic of a traditional social institution of mutual cooperation, the system of minor irrigation is imbedded into the social framework of the rural society of arid zone. Various socio-economic and cultural factors like kinship relationship, caste-based joint ownership of land-water resource, sharing principle, norms towards water use, distribution and maintenance of the irrigation source and emergence of new settlement pattern have characterized its solidarity and continuance for generations. Predominance of the joint family system helps in pooling the human resources for efficient utilisation of the land and water. But with the changing times the system has lost much of its social significance. Individualistic competitive economic pattern resulting in non-homogeneity in relations due to severe socio-economic and physical hazards as an impact of frequent droughts are substituted for jointly and cooperatively owned enterprises.

The various impediments occurring in the system are rigidity of ownership rights to even fractional pieces of land, minor proportion of ownership in irrigation source, minimum subsistence level acceptable, gradual fragmentation of land and water resource, etc., inhibit irrigated cultivation, consequently socio-economic development in the area. Adherence to the fixed social norms of the system leads to a very restrictive resource manoeuvrability and uncertainty in irrigation for the farmers particularly of arid regions. To overcome this there is need for the design of a new or the modification of the existing social norms of the system which will ensure an effective combination of resources and people in terms of land-water relationship.

Various studies deal with the economics of Indian irrigation development (Gadgil, 1948; Raj, 1960; ICAR, 1961; Epstein, 1962; Ansari, 1963; Clark, 1964; Moorti, 1970); impact of the major irrigation system of Rajasthan and the Gang Canal on the social and economic aspects of the population inhabiting the arid zone of Rajasthan (Bharara et al. 1973, 1974; Bharara, 1977) and on institutional rationing of the canal water in northern India (Reidinger, 1974).[5] But almost no study has so far been done on the social context of the minor irrigation system in the arid zone of Rajasthan.

The present study based on empirical data deals with the type of a minor irrigation system and its socio-economic characteristics, sourcewise irrigation potential available, ownership coverage and distribution pattern, control of water use, cultivation and irrigation turns, and the resultant policy implications to safeguard against irrigation insecurity due to drought.

Design of Study

The study was conducted in an arid village Daijar located in *Panchayat Samiti* Mandore of district Jodhpur, Rajasthan. A study of all the sources of irrigation prevalent in the village was made in 1973-74 and supplemented with the field data collected in 1979. The heads of households owning the irrigation sources at different scattered farms of the village were enumerated. A schedule was administered and data were collected by personal interviews with the heads of households. Representatives of various caste groups owning the sources were interviewed. Secondary data pertaining to the size of land holdings were collected from official records. A detailed case study of one of the existing sources giving ownership coverage and its distribution pattern was also made.

Types of Irrigation System

The system of irrigation has been changing. Changes occur in society, as the system develops (Stamp, 1964). In this region the most modern and efficient system of energised wells (power-driven pump installed deep within wells) was evolved through the traditional, less efficient mode of irrigation system of *Charsa* (leather bucket for lifting water) to *Rahat* (Persian wheel) to the diesel engine-operated water pump. This shift from the traditional *Charsa* system to the largely energised well system vividly indicates a steady economic uplift of the farming community and its responsiveness to innovated technology. The duration of the transitional phases depended heavily on the economics and available technology, upon the changes in farming, land tenure system, labour situation, willingness and attitude for capital investment, land improvement and increasing needs of the population.

A wide range in the types of irrigation sources was found (Greene, 1966). Historical trace-up of the system revealed that *Charsa* being the most ancient system of irrigation was still prevalent in the village. It was only after 1955 when the *Rahat* system came into existence. Recently during the good rainfall year of 1970 when power became available, the *Rahat* system developed into a most efficient energised well system of irrigation.

The village had 46 minor sources of irrigation out of which only 33 were in operation during the year 1977-78 (Table 10.1). Out of 13 sources nine consisting of *Charsa* (1), *Rahat* (1) and diesel pump-operated wells (7) had recently (1977-78) been dug out but were not in operation.

Most of the wells are concentrated in Daijar Nali (a natural stream course) followed by sparse locations in scattered settlements of the village. The location has exploration of wells for irrigation were mainly based on the recommendations of the village saints/water diviners rather than those of experts. As a result of digging the maximum number of wells at one place (14 within an area of hardly 1 sq.km. of Daijar Nali) created uneven disturbances in the water resource potential of the wells.

Ownership and Distribution Pattern

Earlier a joint ownership system of minor irrigation had mostly been predominant. The mean number of households per operating source varied from a minimum of 2.3 in the case of *Rahat* to the maximum of 3.0, 3.3 and 7.0 in the case of *Charsa*, energised well, and *Jharna*[6] respectively. On an average a source was owned by three households but used by 4.6 households.

TABLE 10.1: Types of irrigation systems and their users in village Daijar

Irrigation	Number of sources		Number of households		Mean No. of households/operating source	
	Operating	Non-operating	Owning	Using	Owning	Using
Charsa (bucket system)	3	4	10	15	3.3	5.0
Rahat (Persian wheel)	6	1	14	18	2.3	3.0
Well (energised)	23	8	66	112	3.0	5.0
Jharna (natural spring)	1	—	7	7	7.0	7.0
Total	33	13	97	152	3.0	4.6

The extent of ownership proportion in an irrigation source varied from a minimum of 0.02 held by 6.2 per cent of the total irrigated farm owners to the maximum of 1, i.e. individual ownership rights held by 14.4 per cent of these owners (Table 10.2).

TABLE 10.2: Proportion of ownership in irrigation sources in village Daijar

Proportion	Households (No.)	Per cent
1.00	14	14.4
0.50	10	10.3
0.33	6	6.2
0.25	17	17.5
0.17	6	6.2
0.11	7	7.2
0.08	1	1.0
0.06	9	9.3
0.04	9	9.3
0.03	12	12.4
0.02	6	6.2
Total	97	100.0

The highest percentage of the households had an ownership proportion of 0.25 per source. There was too much concentration of small ownership rights on most of the sources as about 70 per cent of the total households had four or more shares per source. Now with the increase in the competitive economic era, individual ownership has increased. Ten out of thirteen non-operating sources dug up during 1978-79 have individual ownership rights.

Joint ownership of most of the sources involved members of the same caste and of the same clan.[7] No relationship was found between the type of caste and the type of well owned. However, a dominant caste, Mali, had the highest number of sources, Table 10.3 shows the castewise distribution of households owning the wells.

TABLE 10.3: Castewise distribution of households owning wells in village Daijar

Caste	Irrigation No.	Wells %	Households No.	Households %	Average number of households per source
Mali	21	63.6	76	78.3	3.6
Brahmin	3	9.1	2	2.1	0.7
Jain	2	6.1	1	1.0	0.5
Purohit	2	6.1	12	12.4	6.0
Jat	2	6.1	2	2.1	1.0
Swami	1	3.0	1	1.0	1.0
Meghwal	1	3.0	2	2.1	2.0
Rajput	1	3.0	1	1.0	1.0
Total	33	100.0	97	100.0	2.9

On an average, three households owned the same source of irrigation. Malis, owning the highest percentage (63.6) of irrigation sources, had the highest percentage (78.3) of households that use irrigation, but the average number of households per source was 3.6 only and the ownership ranging from 1 to 20 households/source. Purohits, on the other hand, had the highest average number (6) of households per source. The households of Brahmins and Jains owned more than one source; these owners were not the original settlers but had influxed from outside and had purchased irrigated farms. Jat, Swami and Rajput households had individual (single) ownership rights and had one source each.

Distribution of the size of land under irrigation and of the ownership share in the irrigation source (Table 10.4) revealed a positive correlation (+0.549) (Reidhinger, 1974). With an increase in the ownership share in the irrigation source, the size of land under irrigation increased. The minimum ownership share of 1/48th in the irrigation source could provide irrigation for hardly one *bigha* (0.16 hectare) of land while the maximum share of one (individual ownership) could provide irrigation up to 80 bighas (12.8 hectares) of land under the source.

About half of the total irrigated farm-owning households had less than five bighas (0.8 hectare) of land under irrigation. Ownership of 5 to 10 bighas (0.8–1.6 hectare) and 10 and above bighas/hectares of irrigated land categories belonged to about 25 per cent each of such households.

Sourcewise Irrigation Available

Underground water is essentially the rainwater accumulated at different depths, due to development of joints and cracks in the hard pan due to weathering.[8] Its extent differed from system to system. The data on systemwise irrigation, as shown in (Table 10.5), reveals that the gradual development of the system brought an increase in the available irrigation. The average irrigation available per system was minimum (1.8 hectare) in the case of *Charsa* and maximum (5.3 hectare) in the case of energised well. On an average, the culturable area and irrigation available per system came to 10.8 and 3.9 hectare respectively during the year. Most of the areas under the different systems were not levelled properly and this resulted in appreciable wastage of water. In such cases education and persuasion of the farmers was needed to level their fields to avoid maldistribution of water.

The differences in the availability of irrigation per source were mainly due to inadequate amount and uneven distribution of rainfall (ranging from less than 180 mm in a bad year to more than 550 mm in a good year), source depth (20 to 60 m), kind of operating machine (bucket to lifting pumps), the location of wells (*nali* bed, plain or hilly area in the same belt or in scattered settlements), the catchment area, recharging of the source, desilting from time to time, etc.

Based on data on rainfall and catchment area, prediction of the available water for irrigation would ensure more effective water use and land match by the farmers. This knowledge in advance would help the farmer to plan and manage the farm. Extension methods can propagate this knowledge pertaining to season and for the system as a whole for the area.

TABLE 10.4: Distribution of ownership share in irrigation source and size of land irrigated in village Daijar

Irrigated land (*bighas*)*	Ownership Share											Total	Per cent
	1/48	1/36	1/24	1/18	1/12	1/9	1/6	1/4	1/3	1/2	1/1		
<1	6	—	9	3	—	4	—	—	—	—	—	22	22.7
1—2	—	—	—	—	1	—	2	—	—	—	—	3	3.1
2—3	—	6	—	—	—	—	—	8	1	2	—	17	17.5
3—4	—	—	—	6	—	—	—	—	—	—	—	6	6.2
5—10	—	6	—	—	—	3	2	5	3	4	1	24	24.7
10—15	—	—	—	—	—	—	2	—	—	—	4	6	6.2
20—25	—	—	—	—	—	—	—	4	2	2	5	13	13.4
25—30	—	—	—	—	—	—	—	—	—	—	1	1	1.0
30—35	—	—	—	—	—	—	—	—	—	2	—	2	2.1
40—45	—	—	—	—	—	—	—	—	—	—	2	2	2.1
75—80	—	—	—	—	—	—	—	—	—	—	1	1	1.0
Total	6	12	9	9	1	7	6	17	6	10	14	97	100.00
per cent	6.2	12.3	9.3	9.3	1.0	7.2	6.2	17.5	6.2	10.3	14.4	100.00	

* The bigha was taken as the land unit, as the extent of land irrigated was too small to be converted into hectares.
1 bigha = 0.1618 ha.

TABLE 10.5: Sourcewise culturable area and irrigated area in village Daijar

Irrigation source	No.	Total CA (ha)	Irrigated area (ha)	% if total irrigated area	Average C.A./ source (ha)	Average irrigated area/ source(ha)
Charsa (leather bucket)	3	27.50	5.33	4.1	9.17	1.78
Rahat (persian wheel)	6	56.33	18.16	14.0	9.39	3.03
Well (energised)	23	270.83	105.17	81.1	11.77	5.25
Jharna	1	2.66	1.00	0.8	2.66	1.00
All sources	33	357.32	129.66	100.0	10,83	3.93

Operational Pattern of an Energised Source

Gradual disintegration of joint into nuclear households had brought continuous division of land-water resource ownership. Its extent is revealed through ownership coverage and distribution pattern of an energised source of irrigation. Typically, this minor irrigation source is being operated by the owners who are each responsible for water control on a specified area on a specified day. Each household owner makes use of the source according to his fixed turn on the fixed day. But it is observed that the farmer/user tends to over-irrigate if he is given the chance, because of the fear of uncertainty inherent in the system. The extent of use of the source is adjusted according to weather conditions and periodic modifications of the water flow being made as and when needed by the users. The possibility of shortage and surplus of irrigation persists in bad and good rainfall years. Social penalties are imposed for contravention of the fixed norms and conventions for the use of irrigation water. Share and right of the water sources are largely determined by the extent of land owned under the source; the number of families disintegrated from joint to nuclear households; kinship relationship; succession of property inherited and sharing capital investment on the continual operation of the source.

Twenty households constituting 10 each of two different clans of the same caste group of the Malis are sharing the same source of irrigation. The allocation of water in proportion to their share in size of land and source of irrigation is made through a two-fold distribution system of two water channels functioning at a time. Each household gets its turn after three days. Out of 20 households owning the source, only six households can do irrigated cultivation in one year and others have to wait for their turn. Some of the households keep on waiting in order to avail of their turn for as long as six years.

The general scarcity of irrigation potential and the extent of ownership of the source determine a household's turn for irrigation at the well (Table 10.6).

TABLE 10.6: Cultivation and irrigation on an energised source

Households getting turn	Period of turn for cultivation	Households availing irrigation/year
1	Every year	1
1	2 years continuously and	1
2	after 2 years	
6	after 3 years	2
4	after 4 years	1
6	after 6 years	1
20		6

The data in Table 10.6 would reveal that out of 20 households, six do cultivation every year while for using irrigation every household gets its turn after three days. Only two households avail of irrigation on the same day from the same source but from two different channels meant for the users. To ensure that each household gets its turn for irrigation, a minimum of six years is required for the functioning of the system. Within six years a combination of six households in each year will operate upon the source. These combinations (Fig. 10.1), worked out according to their ownership share, are (i) $a_1\ b_1\ c_1\ x_1\ y_1\ z_1$; (ii) $a_2\ b_2\ c_2\ x_1\ y_2\ z_2$; (iii) $a_3\ b_3\ d_1\ x_1\ y_3\ z_3$; (iv) $a_4\ b_1\ d_1\ x_1\ y_4\ b_4$; (v) $a_1\ b_2\ c_1\ x_1\ y_2\ b_5$; (vi) $a_2\ b_3\ c_2\ x_1\ y_3\ b_6$.

Within a span of six years, eight households ($a_3\ a_4\ z_1\ z_2\ z_3\ z_4\ z_5\ z_6$) will get one turn each; 11 households ($a_1\ a_2\ b_1\ b_2\ b_3\ c_1\ c_2\ d_1\ y_1\ y_3$) will get two turns each; and one household (x_1) will get six turns and all households in these categories will get benefit/disadvantage once, twice and six times respectively.

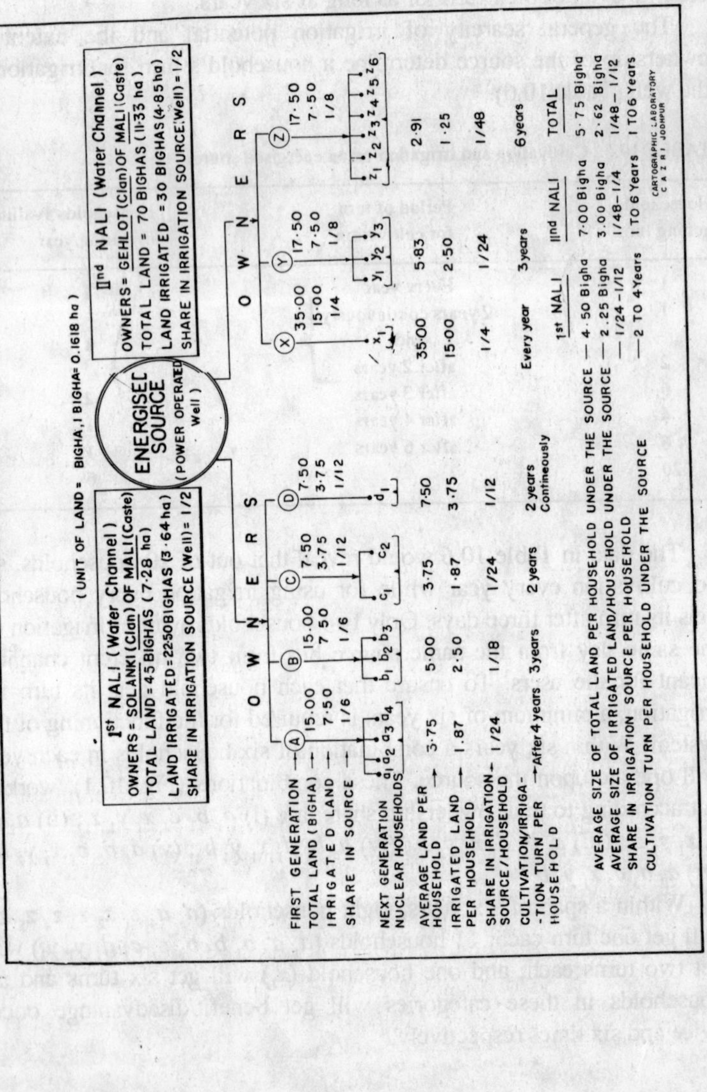

FIG. 10.1: Distribution of land-water resource ownership in an irrigation source in village Daijar (1973-78)

In the arid zone of Rajasthan, almost every third year is a famine year, so in a period of six years, two years are considered as famine years (bad years, 180 mm rainfall), three years as average (normal years, 180-550 mm rainfall), and one year as very good (above normal, i.e. 550 mm and above rainfall). The underground water available to the farmers during famine periods is almost negligible, during average periods it is just normal and during good years it is almost surplus. The cultivators whose turn comes in the bad years are exposed to more risk and uncertainty in agriculture.

To cover such risk and uncertainty, there is genuine need for establishing adequate and workable social security measures. Ideally, the households availing mostly good years should contribute towards risk compensation in proportion to their ownership share in land and irrigation source.

Policy Implications—Suggestions

Empirical studies revealed numerous social and economic problems, viz. excessive overcrowding of irrigation source, fragmentation of ownership of land irrigation resource, collective responsibility towards maintenance, management and distribution of water resource, uncertainties, lack of technical know-how, limited irrigation potentials, non-scientific exploration and location of well sites, risk and uncertainty in irrigation potentials leading to crop failure, etc.[9] These problems stimulated the growth of local organisations for risk minimisation and compensation as a measure of social security against the drought periods. Such an association like, "the Farmers' Association for Rural Water Resource Use" can go a long way in maintaining continuity and solidarity of the system making it a risk-relieved and successful socio-economic system. A network of such associations spread over the whole of the arid region will safeguard the farmers against drought through institutionalised cooperation.

Most of the socio-economic problems accruing to the system need government efforts through legislation and self-participation programmes of the villagers. Some of the suggestions, include:

(i) assessment of land and water resource potentials through surveys. This will reduce the risk and uncertainty in the selection of sites for wells which is being done now purely on the recommendations of the local water diviners,

(ii) investigations of techniques of water control by evolving flexible cropping patterns best suited to bad, average and good rainfall years. Researches on determining effective irrigation used by different cropping patterns and on the size of the optimally productive irrigated unit under three different rainfall regimes will help in formulating suitable recommendations,

(iii) education of all concerned in irrigation methods, land and water conservation,

(iv) planning overall development of land and water so as to optimise its utilisation and consequently avoid wastage, and

(v) assisting private joint owners by means of loans and grants for energising their sources.

Conclusion

A study of the well-irrigation system in the arid zone of Rajasthan revealed a preponderance of joint land-water resource ownership in the region. Too much concentration of households and small ownership rights on most of the sources of irrigation was prevalent. The size of land under irrigation was related to ownership share in the irrigation source. Village water diviners still continue to play a significant role in the selection of sites for wells. The system of irrigation has been changing. The gradual shift from the traditional *Charsa* system to the largely energised well system vividly indicates a steady economic uplift of the farming community and its receptiveness to innovated technology.

The distribution of water has been traditionally based on ownership shares. This has led to a very restrictive resource monoeuvrability and uncertainty for the farmers whose turns fall during poor rainfall years. To cover such risks and to provide compensation to the farmers who would have their turn of water use in the drought years, the study suggests introduction of a social security system through a farmers' association which would raise contribution proportionate to water use shares, from those farmers who happen to be fortunate to use water during good rainfall years.

NOTES

1. K. L. Rao (1975) states that since independence the country has spent Rs. 8,400

crores on irrigation, out of which Rs. 2,800 crores alone were spent in digging wells and tanks. (*India's Water Wealth*, New Delhi, Orient Longman, 1975).

2. *India 1975* (p. 201) records that between 1968-69 and 1973-74, the number of dugwells increased from 5.7 million to 6.5 million electrical pumpsets from 1.1 million to 2.4 million, private tube-wells from .266 million to .82 million and public tube-wells from 15,000 to 20,000. The Fourth Plan Target of covering 7.2 million hectares by minor irrigation has been exceeded by about 10 lakh hectares. The total irrigated area in the country, which was about 2.26 lakh hectares in 1950-51, rose to about 44.7 million hectares in 1973-74. Under the minor irrigation system, it increased from 12.6 million hectare in 1950-51 to 18.7 million hectares in 1970-71 (Govt. of India, Publications Division, Ministry of Information and Broadcasting, pp. 200, 163)
3. The *Report of the Irrigation Commission*, Vol. II (Calcutta: Govt. of India, Ministry of Irrigation and Power, Sree Saraswati Press, Ltd., 1972) pp. 345-6, refer to increase in the irrigated area under wells and in the number of wells in Rajasthan. Of the 2.119 million hectares irrigated area in Rajasthan, 1.176 million hectares, i.e. more than half, was commanded by wells during 1965-66. The number of wells in use rose from 0.45 million in 1961-62 to 0.61 million in 1968 and the area irrigated has risen by 1 million hectare in 1965.
4. Since the beginning (1970-71) of the Drought Prone Areas Programme (DPAP) stress has been laid on better planning, investigation and development of ground water resources in the drought-affected regions. Minor irrigation schemes accounted for 60 per cent of the total sanctioned budget of the DPAP schemes in 1970-71.
5. The study reveals that rationing of canal water introduces uncertainty in farmers, water supply in addition to scarcity of water.
6. Jharna: (natural spring) is a continuous side seepage of water impounded in inter-hillock area of the village.
7. Further elaboration makes it clear that the system of operation of wells by several cultivators jointly was termed as *Bhawli* which cuts across caste and kinship boundaries (Central Arid Zone Research Institute Jodhpur, Land Transformation Plan of Lune Block in Jodhpur district, Western Rajasthan, Divisional Cyclostyled Report No. 1969/1) p. 110.
8. These results were obtained through seismic survey by the Rajasthan State Underground Water Department while exploring possibilities of irrigation potential available in the village.
9. James and Lee (1971) emphasised formulation of Water Resource Planning Acts and Ground Water Laws by various countries and their appropriate application in arid tracts. Similarly, a land-owner should be granted the right to pump ground water at any point on his property as long as he does not waste or sell water and does not act maliciously. Permission to drill a new well should be granted by an expert agency provided sufficient water is available and no serious detrimental effect will result to older wells. Thus formulation and legalisation of such type of government actions, as suggested by the author, will go a long way in making a permanently successful socio-economic minor irrigation system of the arid regions. Introduction of a social insurance scheme of assured irrigation compensation lessening economic decay will give practical shape to the concept of social welfare and raise productivity.

REFERENCES

Ansari, N., 1963 : Some Economic Aspects of Irrigation Developments in N. India, *J. of Agrarian Affairs*, p. 40.

Bharara, L. P., et al., 1973 : "Impact of Irrigation on the Changes in Population, Characteristics of Beneficiary of Non-beneficiary in a Desert Region: *Indian J. of Social Research*, Vol. XIV, No. 2 (1973), pp. 137-45.

——— et al., 1974 : Some Socio-agricultural Changes as a result of Introduction of Irrigation in a Desert Region, *Annals of Arid Zone*, Vol. 13, (March 1974), pp. 1-10.

———, 1977 : "Socio-economic consequences of canal irrigation in an arid district of Rajasthan", *Annals of Arid Zone*, Vol. 16 (4), 1977, pp. 458-72.

Clark, C., 1964 : "The Economics of Irrigation," *Indian J. of Agri. Econ.*, pp. 25-32.

Epstein, T. S., 1964 : *Economic Development and Social Change in South India* (USA : Manchester Univ. Press).

Gadgil, D. R., 1948 : "Economic Benefits of Irrigation," *Report of Survey of the Direct and Indirect benefits of the Godawari and Pravada Canals* (Gokhale Instt. Pub. 17).

Greene, H., 1966 : "Irrigation in Arid Lands," in *Arid Lands*, ed. E.S. Hills (London: Methuen & Co. Ltd., Paris : UNESCO, 1966), p. 255.

ICAR, Crop Patterns for the Bhakra Canal Area, *ICAR Bull.* No. 87, 1961.

James, L. D., and Lee, R.R., 1971 : *Economics of Water Resources Planning*, New Delhi, pp. 151-52.

Moorti, T. V., 1970 : A comparative study of well irrigation of Aligarh District, India (Ithaca, N.Y. : Cornell University, Department of Agri. Eco., Occasional Paper No. 29).

Raj, K. N. : *Economic Aspects of the Bhakra Nangal Project*, Asia Pub. Co., New Delhi.

Rao, K. L., 1975 : *India's Water Wealth*.

Reidinger, R. B., 1974 : "Institutional Rationing of Canal Water in Northern India : Conflict between Traditional Patterns and Modern Needs," *Economic Development and Cultural Change*, 23 (Oct. 1974), pp. 79-104.

Stamp, L. D., 1964 : *A History of Landuse in Arid Regions* (Paris : UNESCO Arid Zone Ser., Vol. 17, 1964), p. 388.

Steward, J. H., 1955 : *Irrigation Civilisation : A Comparative Study* (ed.) Washington, D.C.: Pan American Union Soc. Sci. Mongr. I, 1955, p. 78.

11

FARM DIVERSIFICATION AND THE RESTRUCTURING OF AGRICULTURE

BRIAN W. ILBERY

MANY WESTERN economics have undergone second agricultural revolution in the post-war period (Heady and Ilbery, 1985). Farming has been restructured, a process characterised by a concentration of capital and output on fewer, larger farm businesses, and an increasing specialisation and intensification of production (Boiler, 1966; Marsden et.al. 1986 and Sargent, 1986). Among the consequences of such modernisation are a massive rise in food production (the European Community (EEC), for example, is now more than self-sufficient in 16 agricultural products): an increasing polarisation in farm incomes; an increasing dependence of family farms on industries upstream and downstream of farming itself; and adverse environmental effects on the rural landscape.

Agricultural modernisation was deliberately sponsored by policy makers, who attempted to solve the social problem of low farm incomes with improved technology and a range of price support measures and capital grants. The relatively risk-free economic environment created by the Common Agricultural Policy (CAP), for example, served only to exacerbate the growing problem of overproduction. Paradoxically, as price support levels have risen, farm incomes have fallen (by an average 40 per cent in the United Kingdom between 1984 and 1986 alone). This unexpected situation reflects a treadmill (Dexter, 1977) upon which farmers have been placed, where inflation and a decline in the demand for food relative to income levels have caused the costs of production to increase at a greater rate than the price of foodstuffs (price-cost squeeze). In an attempt to maintain or increase incomes, farmers sought to obtain economies of scale, either through expanding their land area or by increasing their intensity and/or specialisation of production. Government policy was again instrumental and farmers increased their borrowing of capital. This inevitably increased levels of indebtedness and the only way to survive was to intensify production even further (Sargent, 1986). The resultant overproduction had adverse effects upon prices, and thus farm incomes, and a vicious circle had developed.

Although partly responsible for the high levels of farm debt, many government in the western world began to change their agricultural policies in the 1980s. Recognising, at last, the need to curb the ever-increasing problem of surplus food production, the freedom of farmers to continuously expand production began to be constrained. In the EEC, this has taken the form of milk quotas (1984); further agreements on milk and beef (1986); a restrictive pricing policy for certain products; and, most recently (June 1987), a Regulation concerned with the 'extensification' of production, whereby farmers will be compensated for reducing their area of particular crops and number of livestock by 20 per cent.

It is becoming increasingly clear, therefore, that farmers will not be able to respond to low incomes and indebtedness by either maximising production or transferring to non-quota products. Understandably, many farmers, especially those with small farms, feel bitter and hold the government responsible for their precarious position. However, the need to adapt to market conditions and give due consideration to the rural environment has now been recognised by policy makers and farmers alike. A further restructuring of farm businesses is necessary and, while various possibilities present themselves, the most commonly discussed option is the development of 'alternative enterprises' - generally known as farm diversification.

Farm Diversification

Much has already been written on the economic viability and problems of different types of farm diversification. However, the term has rarely been adequately defined. According to the National Farmers Union in the UK, farm diversification covers a 'multiplicity of situations, which can often only be adequately defined as doing different'. These range from an adjustment of inputs to produce conventional products in a different way (organic farming) and switching enterprises into new areas where demand is currently met by imports (for example, snails, herbs), to developing value-added processing and marketing of produce and introducing new enterprises outside food production (for example, farm tourism and recreation). Off-farm sources of income, from other economic activities, are not included, although it is recognised that they may provide the necessary capital for on-farm diversification (Gasson, 1983).

Diversification into alternative enterprises, therefore, implies a contrast with conventional enterprises. The Centre for Agricultural Strategy presented and discussed a long list of alternative crop and livestock enterprises; this excluded value-added alternatives. In contrast. Griffiths (1987) emphasised their importance, but excluded alternative crops and livestock! He defined farm diversification as 'farm-based activities not directly concerned with producing crops or livestock and which involve marketing contact outside the agricultural industry'. Consequently, value-added activities (by processing and marketing), farm tourism, and recreation services formed his main categories of diversification.

The most all-encompassing definition of farm diversification is that given by Slee (1987).

"Those enterprises taking place on predominantly agricultural proprietal units which: (a) are not based on the primary production of food and fibre and/or (b) fall outside the price support mechanisms of the CAP."

This gives rise to five main groups of activities, which are schematically portrayed in Fig. 11.1 – tourism, recreation, adding value to conventional products, unconventional enterprises, and the use of ancillary buildings and resources. Tourism and recreation are often combined into one group and the last group (ancillary buildings etc.) is sometimes incorporated into the other categories. As Fig. 11.1 demonstrates, the groups are not mutually exclusive, as redundant farm buildings, for example, can provide the input material for holiday accommodation. Nevertheless, Fig. 11.1 is fairly comprehensive and provides a useful framework for comparative empirical analysis.

Factors Affecting farm Diversification

Many farmers will begin to diversify their businesses as a matter of economic necessity. However, this provides only a simplistic explanation of the development of farm diversification and other interrelated factors are necessarily involved. These range from macroscale factors to specific features of individual farms and farmers and include:

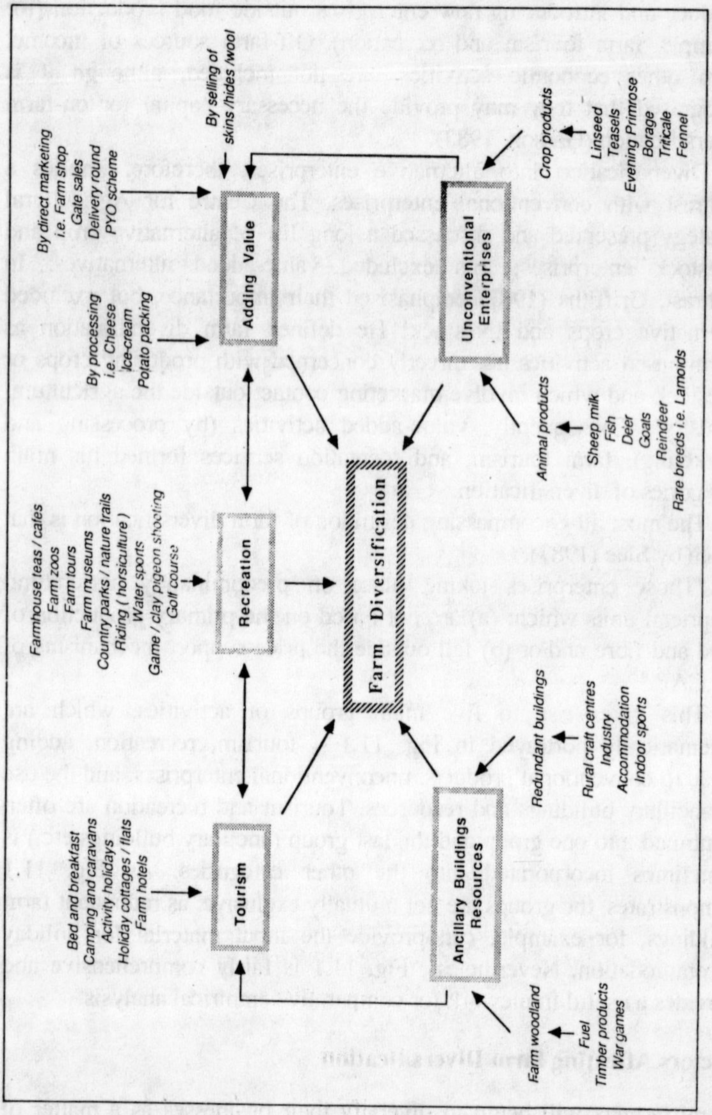

FIG. 11.1: Schematic flow diagram of the major types of farm diversification.

The development of policy in agriculture and by other rural bodies. In addition to increasing EEC pressures to diversify, individual countries are introducing their own solutions. The British government's (1987) ALURE package (Alternative Landuses for the Rural Economy), for example, offers £ 25,000,000 per year towards the development of farm woodland, forestry, diversification, and Environmentally Sensitive Areas. It advocates a relaxation in rural planning controls, except on high quality land and in Green Belt areas. This is 'in line' with other policies for job creation and the conversion of redundant farm buildings in rural areas produced by the Development Commission and COSIRA (Council for Small Industry in Rural Areas), and for tourist and recreational potential in the countryside (such as Tourist Boards, Sports Council). Farm diversification is, therefore, being pushed by non-agricultural bodies and agencies.

The changing nature of the rural population in the 1980s. Counter-urbanisation and the urban-rural shift of industry are creating new potential markets for farmers. The movement of people to rural areas has been dominated by the higher socio-economic classes, who place great importance on environmental quality and have the means to pay for farm-fresh products and rural recreation.

The growing national trend towards the conservation of the rural environment. Pressure groups are beginning to influence policy decisions and have helped to promote a wider interest in healthy, often organic, food. Specialist markets for both new and traditional products are emerging and farmers are well-placed to exploit such opportunities.

The growing media attractiveness of diversification. Diversification is at present a 'buzz word', reflecting the promotion activities of various organisations with a vested interest in it. Farmers will find it difficult not to read about alternative farm enterprises and gradually attitudes will change as they become won over to the ideas of agricultural adjustment.

The need to provide employment for member(s) of the farm household. Many alternatives are started as small scale ventures, often by farmers' wives (for example bed and breakfast, farm catering). However, it is now being realised that farm diversification offers considerable potential for involving all family members, as well as putting the farm business on a more viable footing.

The type of farmer and location of the farm. Some farmers (innovators) have better business skills than others and are able to recognise and take advantage of an unsatisfied demand. This is especially so if the farm is within easy reach of a large concentration of people (the urban fringe) or located on major transport routes and in popular tourist areas.

The success of other farmers' activities. As with the diffusion of other agricultural innovations, many farmers will be influenced by what they see over the fence: there are, however, certain problems associated with such an approach (see later).

The Development of farm Diversification on the Urban Fringe

It is reasonable to assume that the factors affecting farm diversification will have a spatially-varied impact, with some areas offering considerably more potential than others. The urban fringe (a 15 to 30 km zone of land surrounding cities) provides one of the best locations, enabling farmers to tap into a large market of potential customers. However, not all farmers on the urban fringe will be equally able to respond to the opportunities and more information on the development of farm diversification is needed. Such a detailed empirical investigation has been conducted for the urban fringe area between Coventry and Birmingham, in England (Griffith, 1987; and Ilbery, 1987). Only a brief outline of some of the findings can be presented in this section. Which concentrates on the following questions:

1. How quickly is farm diversification developing?
2. What types of diversification are involved?
3. What type of farm and farmer are interested?
4. How important, financially, is farm diversification?

Farm diversification has diffused quite rapidly in the area, 73 per cent of the 120 farms with alternative enterprises (1986) having started them since 1975. Only 11 farmers had embarked on farm diversification before 1965, mainly in the form of adding value by direct marketing (farm gate sales and farm shops). Over the next ten years an average 2.2 farms per year began to diversify, a figure that rose to a high of 9.6 for the late-1970s before falling back to 6.5 for the 1980s. Interestingly, pick-your-own schemes (PYO) became important in the late-1970s, replacing farm delivery rounds; similarly, farm tourism has been introduced only since 1975. Proximity to an urban centre is clearly important in the diffusion process, as 64 per cent of the 120 farms are located within 5 km of either Coventry or Birmingham. Indeed, 82 per cent of the farmers stated that their urban fringe location was an important factor in the decision to diversify. However, the increasing national pressures to diversify suggest that the dominance of the urban fringe should be reduced in the future.

Five types of farm diversification were identified in the study. In rank order these were adding value by direct marketing (82 farms); farm-based recreation (37); farm tourism (28); commercial activities (including adding value by processing and craft centres) (24); and other activities (land and building lets, refuse dumping, farm woodland) (9). Direct marketing is, therefore, dominant, taking the form of farm gate sales (54), farm shops (24), PYO schemes (10), and delivery rounds (6). Some farmers have more than one type of direct marketing and, similarly of those involved in such activities, 15 also have tourist enterprises (mainly caravans and bed and breakfast); 20 have recreational activities (dominantly 'horsiculture'); and 12 have commercial activities (Table 11.1). Just over 60 per cent of the farms have just one of the five main diversification types, a similar result to that obtained in a study of the Less Favoured Areas of Britain [14]; only 13 farms could boast three or more types.

TABLE 11.1: Diversification matrix for Birmingham fringe study

	1	2	3	4	5
1. Direct marketing	82				
2. Farm tourism	15	28			
3. Farm recreation	20	6	37		
4. Commercial activities	12	4	4	24	
5. Others	7	3	2	3	9

Source: Author's survey.

Detailed analysis failed to reveal significant statistical associations between the number of types practised and such variables as farm size; farm type; the tenure of the farm; and the age, education, and experience of the farmers. However, farm diversification is biased towards the larger farms in the Coventry and Birmingham area. While 38 per cent of all farms in the area are less than 10 hectare in size (national percentage is 28), only 24 per cent of those with alternative enterprises are under that size. Similarly, diversification was shown to favour farms with extensive livestock enterprises (beef/sheep) and to avoid dairy farms.

Following the work of Marsden et al. [3], the farmers with alternative enterprises appear to be of three different types:

Hobby and retired part-time farmers. These have minimal farm income and no farm debt or mortgage. For example:

A local government worker in Birmingham who purchased a small holding in 1971. The farm is essentially non-profit making, but the owner enjoys the outdoor life after a day in the city. A few sheep, poultry and goats are kept and the farm offers gate sales (of eggs and previously goat milk and cheese) and has a permit for six caravans.

Survivors through diversification. Farmers who have gone into non-traditional agricultural activities as a positive step to increase profits and the scale of business. This is typified by the following:

A 6 hectare livestock farm where the farmer diversified into farm-based tourism and recreation as a way of raising much needed capital to survive and invest in a calf-fattening enterprise. The farm has effectively become a children's farm and zoo, where children are free to play feed the animals and have their packed lunches in the outbuildings. Guided tours, birthday and Christmas parties, and farm activity courses (with accommodation) are available. Caravans and a farm shop are integrated into the whole scheme, with the latter offering eggs, potatoes, sweets, ice-cream and an under 50 p shelf for children(!). Diversification has been so successful that the farmer has postponed the calf enterprise.

Accumulators of capital. Farms where family businesses develop non-agricultural or agriculturally-related investments either to increase returns on surplus capital raised from farming or to use excess capital from non-agricultural businesses to begin in farming. The business remains family-based. A good example of this type of farmer in the study area is:

A 486 hectare mixed farm, with a large poultry unit, where the owner began to diversify through farm gate sales of eggs. This was soon followed by a farm shop and the farmer became the largest egg retailer in the country. Following detailed market research, the shop developed bakery and butchery businesses and customers are able to shop separately in each of the three sections and so avoid unnecessary queues. The farm shop has a turnover of approximately £1 million and employs 22 people on Saturdays; it is complemented by a PYO enterprise in the adjoining field and farm-based recreation (horses), and the family has an interest in three sports centres.

Financially, farm diversification is yet to become a major income earner in the Coventry/Birmingham area. Of the 110 farmers providing income details, only 24 obtain over 50 per cent of their income from such ventures; a majority still earn less than 10 per cent in this way. However, this varies according to the type of diversification, from very few farmers with direct marketing and recreational enterprises earning over a half of their income from such activities to 30 per cent for commercial activities. Interestingly, it is farms that have been losing land which have come to rely most on farm diversification (survivors).

Although the development of farm diversification within the study area has been quite rapid, its uptake is not wholly straightforward and a number of resistances to the natural diffusion process exist. It is to some of these that the rest of this paper is addressed.

Problems involved in the Development of Farm Diversification

Farm diversification is not a panacea for solving the problems of low farm incomes and agricultural over-production Crocker, 1986. The laws of supply and demand dictate the viability and spatial distribution of alternative farm enterprises. A number of considerations have to be taken into account before embarking upon the sort of activities listed in Fig. 11.1. Some of the more general problems are outlined below:

A Change in Farmer Skills

Farm diversification necessitates a change from a production orientation, where levels of output are stressed rather than overall economic efficiency, to one which emphasises marketing skills Slee, 1987. This represents a major transition and the incentive to change from a protected market system (guaranteed prices) to free market competition remains small. Many farmers are unhappy at the prospect of opening their farms to visitors and success will very much depend upon the expertise and business acumen of those involved. Imagination and a positive approach are vital ingredients, although those most in need of diversifying are often the least able to innovate.

Availability of Capital

With falling farm incomes, many farmers do not have the necessary resources to invest in alternative enterprises. Government policy measures are helping to provide capital and marketing grants, but the amount of money available in the ALURE package, for example, is equivalent to less than £50 for every farm in Britain. As with many other agricultural grants, it is only the more able farmers who can plan sufficiently far ahead to satisfy the qualification criteria. With over 98 per cent of the annual UK agricultural subsidy continuing to encourage overproduction, there seems little real government commitment to farm diversification.

Lack of Advice and Research

Less than one-third of the farmers in the Birmingham sample hand received advice before establishing alternative enterprises on their farms. Even less had conducted serious research into the possible options or such issues as market trends, constraints, resources profitability, sources of advice, and legislation procedures. A large number of public and private agencies are now offering advice on farm diversification, but this is serving only to confuse farmers who do not know which way to turn. Greater interagency cooperation and a more coordinated framework for advice are urgently required; otherwise, diversification will continue to develop in a largely *ad hoc* and unresearched manner.

Landlord-tenant Relationships

The urban fringe project demonstrated that over time more tenant farmers are beginning to diversify their farm activities. This will again create problems, because tenancy agreements do not cover activities outside mainstream food supply. Consequently, tenant farmers attempting to introduce farm-based recreation and tourism, for example, may be faced with higher rental charges or given notices to quit. Much depends upon the flexibility of landlord-tenant relationships, which in the study area was shown to vary. Clearly, such relationships warrant proper investigation before proper guidelines can be established for tenant farmers wishing to diversify.

Landuse and Rural Planning

Although the ALURE package advocates an easing of rural planning controls, much still depends upon the different interpretations of the rules by county and district councils. The whole of the Birmingham study area is protected by Green Belt legislation and many farmers have been refused planning permission to erect new buildings or change the use of existing ones (for farm shops, rural crafts, etc). Thus, while the government is encouraging the re-use of redundant farm buildings, they are, in certain areas, also preventing it (the Green Belt, in theory, remains protected under the ALURE package). Rural planners also have to consider the scale of proposed alternatives, together with their effects upon the rural infrastructure and the 'perceived image' of the countryside.

Health and Safety Regulations

Alternative farm enterprises have to satisfy a range of conditions concerned with health regulations (adding value by processing, for example) and safety (insurance for accidents on farm tractor rides, for example). This should be part of an advice and research package, but it has already been remarked how ill-informed many farmers are.

Geographical Location of the Farm

Different types of agricultural area (urban fringe, marginal fringe, prosperous lowland areas) offer different opportunities for farm diversification. Market competition, nationally and within the local area, needs to be properly researched, together with an assessment of potential demand. The actual siting of the farm, in terms of access to transport routes and centres of population, is another important consideration.

Conclusion

Following a post-war restructuring of western agriculture, the socio-economic and environmental context of farming continuous to change rapidly in the 1980s. With farm incomes falling in real terms, farmers are having to consider other sources of income (on and off farm). Some of the factors affecting the diversification process have been outlined in this paper and it seems that farm diversification is still stronger in intent than in mechanisms to make it happen. Only stage one in the process has been reached — the courtship. Stage two — rationalisation — is bound to follow, as the weak fail to survive and the pitfalls of diversification become clearer. As Griffiths (1987) remarked, much more is needed on understanding how diversification has occurred, when, where and why — and why not. This necessitates further theoretical and empirical work, including an examination of farmers' attitudes and motives in relation to their land-use decisions. How do farmers perceive the options open to them and why? How do they feel about agencies outside mainstream agriculture increasingly affecting their decisions? What are their reactions to the new rural population, which is not connected with farming yet expects a certain type of countryside?

The future for farming in the western world is uncertain, yet it is certain that major changes in agricultural land-use will take place before the end of the 20th century.

REFERENCES

Bowler, I. R., 1986 : *Geogr.*, 71, 14.

Centre of Agricultural Strategy, 1986 : Report 11, Reading.

Crocker, S., 1986 : *J. of R.A.S.E.*, 147, 26.

Davies, E. T., 1983 : *Report 218 Agricultural Economic Unit*, Exeter.

Dexter, K., 1977 : *J. Agric. Econ.*, 28, 211.

Gasson, R., 1983 : *'Gainful Occupations of Farm Families'*, Wye.

Griffiths, A., 1987 : *Farmers' Attitudes to Diversification*, Working paper, Exeter.

Healey, M. J. and Ilbery, B.W., 1985 : 'The *Industrialisation of the Countryside*', GeoBooks, Norwich.

Ilbery, B. W., 1987 : *Geogr. Mag.*, 59, 249.

Ilbery, B. W., 1987 : *J. of R.A.S.E.*, 148.

Marsden, T., 1986 : et al. *J. Rural Studies*, 2, 271.

National Farmers Union, 1986 : 'Farm Diversification'.

Sargent, D., 1986 : *Outl. on Agric*, 15, 151.

Slee, B., 1987 : *Alternative Farm Enterprises*, Farming Press.

12

WATER RESOURCE FOR IRRIGATION AND AGRICULTURAL DEVELOPMENT IN MADHYA PRADESH

S. K. SHARMA AND C. K. JAIN

AMONG THE inputs essential for agricultural development water is one. Its use is conditioned by several variables. While low rainfall and its variable nature necessitate the development of artificial means of moisture supply, the increasing employment of fertilisers and, to some extent, of improved variety of seeds make timely supply of water a prerequisite. On the other side, availability of water resource, socio-economic conditions and nature of terrain affect the supply side. These variables cause temporal and spatial variations in the phenomena of irrigation. This paper is an attempt to analyse these variations — temporal and spatial — and to see their correlation with the levels of agricultural development.

Madhya Pradesh presents a diversified picture of land and people. Level land is very limited in some of the river valleys such as the Chhatisgarh basin, the Narmada valley. But most of the part is the assemblage of hills, uplands and plateaux. The Malwa plateau in the western part enclosed by the Madhya Bharat plateau, the Bundelkhand upland, the Vindhyan scarplands and the Vindhyan range presents a typical character of Deccan trap with an undulating surface. The Satpuras, Baghelkhand plateau and the Bastar plateau are tangled masses of hills and plateaux, highly dissected and widely covered with forests. The nature of terrain has manifested itself in the use of water for irrigation and in the levels of agricultural developments (Fig. 12.1).

The distribution of rainfall, mostly controlled by the location in respect of relief, seems to have influenced the development of water resources. Rainfall varies from more than 150 cm. in the south-eastern part to less than 75 cm. in the north-western part of the State. But it is not only the amount of rainfall which conditions irrigation. The high intensity of irrigation in the Chambal valley area may be due to low rainfall but in the Chhatisgarh area including Balaghat, located in a high-rainfall zone, it is the nature of crop and the modern agricultural practice which require surety of water supply.

Present Position

Madhya Pradesh with more than 41.6 million people in 1971 is one of the underdeveloped States. The agriculture, which provides more than half of the value of net domestic products, is still characterised by low productivity, in spite of the formulation and implementation of several programmes during the plan periods. The average yield of foodgrains in the State is only two-thirds of the average for the country.

Natural conditions certainly play a vital role in the performance of agriculture; human factors such as use of inputs also determine the agricultural productivity. The position of this State in respect of the use of inputs is far behind the national average. For instance, only 9.8 per cent of the total cropped area is irrigated (lowest among the States except Tripura) while the country's average was 25.4 per cent in 1974-75. Similarly, in adoption of other development measures the position of the State is not much better. It is presumed that basic among the development measures necessary for the transformation of the face of agriculture in this State is irrigation, others are very closely associated with it. The further analysis intends to see the relationship between temporal and spatial of irrigation and other developmental measures carried in this State.

Spatial Pattern of Irrigated Area

The proportion of irrigated area varies markedly in different parts of the State as exhibited in Fig. 12.2. It ranges between almost nil in Konta tehsil of Bastar district and 48.1 per cent of total irrigated area in Dabra tehsil of Gwalior and 51.0 per cent in Waraseoni tehsil of Balaghat. In all 60 out of 190 tehsils record higher percentages than the average (9.8%) for the State. Among them, two tracts with comparatively high and very high percentages come out very clearly. One is the Chhatisgarh basin encompassing most of the tehsils of Raipur, Bilaspur, Durg and also Balaghat districts. It is to be noted that it is the "rice bowl" of the State and rainfall is generally more than 120 cm. The second tract of a highly irrigated area extends over Gwalior, Bhind, Morena, Shivpuri, Datia, Tikamgarh and Chhatarpur districts. These districts with markedly low rainfall predominate in such Rabi crops as wheat and jowar. Besides, six out of eight tehsils of Mandsaur have a higher proportion. Only a few tehsils in western Malwa and in the lower Narmada valley disrupt the monotony of least irrigated tract. Most often the proportion of irrigated area in these tehsils is between 100 and 150 per cent of the State average.

FIG 12.1: **Madhya Pradesh (1976).**

Conversely, the eastern part of the State excepting Chhatisgarh is least irrigated. In Bastar (1.2%), Surjuga (2.1%), Mandla (1.1%), Sidhi (1.8%) and Shahdol (1.2%) irrigation has not started in the real sense. In the western part, Vidisha is the only district. In short, the Bastar plateau, Baghelkhand plateau, Maikal plateau, Rewa plateau and the Vidisha plateau fail even to employ such basic input as irrigation. Nevertheless, the western part is nearer to the State average than the eastern part.

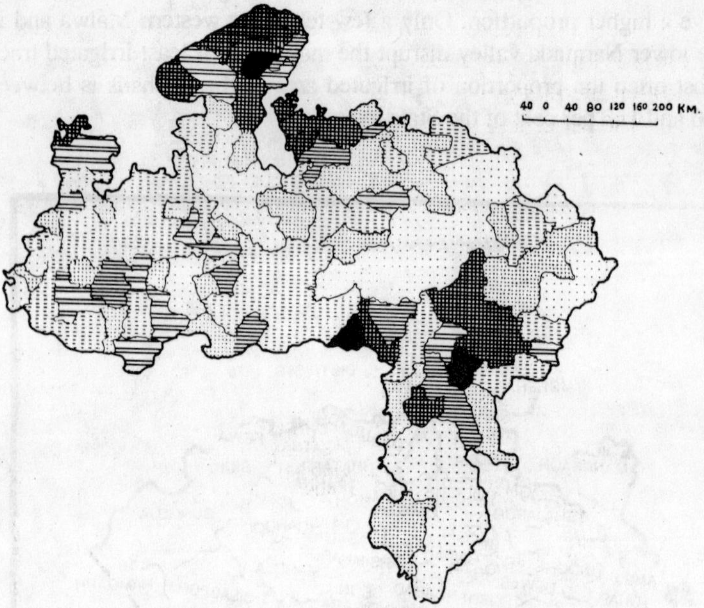

FIG. 12.2: Madhya Pradesh: Proportion of irrigated area, 1976-77.

There seems to be a direct correlation between irrigated area and nature of terrain and thereby fertility of soils. By virtue of comparatively level lands and fertile fine-grained loamy soils in the Chhatisgarh, parts of the Madhya Bharat plateau and Bundelkhand uplands, it is not only easy but economic also to develop means of irrigation. It is here that the investment of the vast capital is capable to pay returns by facilitating the expansion of cropped area, intensification and by substituting crops by more remunerative crops. Other resources required for agriculturisation and further commercialisation of agriculture are more readily available in these easily accessible areas. The relationship between these development measures and development of irrigation has been examined in detail later.

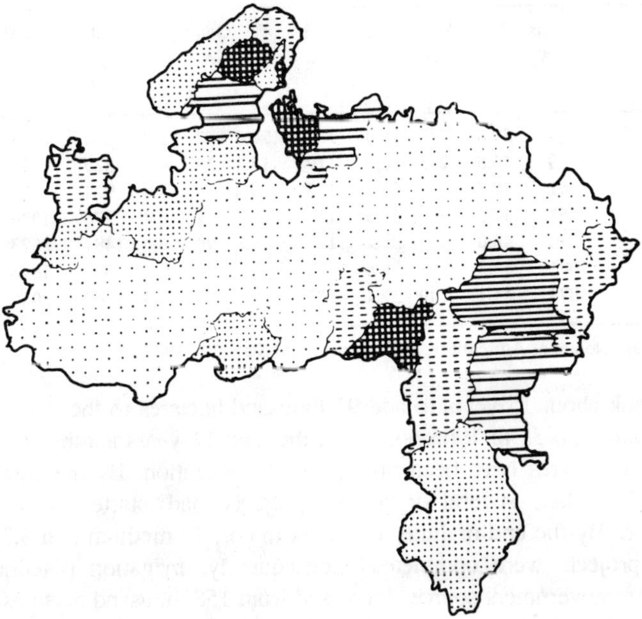

FIG. 12.3: Proportion of irrigated area, 1950-51.

Growth of Irrigation

Irrigation has been practised quite a long time in a few parts of the State. The Gwalior region of Madhya Bharat and the Chhatisgarh region of C.P. & Berar are worth mentioning, where irrigation potentials were created during 1920–40. But in other parts, development of irrigation is the creation of the Plans. In the beginning of the First Five Year Plan, 1950-51, gross irrigated area was 905 thousand hectares which was only 5.9 per cent of gross cropped area (Fig. 12.3). During the past 26 years of Plan regime several medium and minor projects and a few major projects have been executed and completed. As a result of these efforts, total irrigated area increased to 2,050.3 thousand hectares in 1976-77, constituting 9.8 per cent of the cropped area and recording an increase of 126.4 per cent. The progress of irrigated area is furnished in Table 12.1.

TABLE 12.1: Progress of irrigation in M.P., 1950-77

('000 hect.)

Year	Canals A	%	Wells A	%	Tanks A	%	Other A	%	Net irrigated A	Total Gross A
1950-51	354	39.8	249	28.0	235	26.4	52	5.8	890	905
1955-56	370	44.8	298	36.0	119	14.4	38	4.8	825	837
1960-61	441	47.7	324	35.1	122	13.2	37	4.0	924	938
1965-66	461	47.2	345	35.3	118	12.1	52	5.4	976	1000
1970-71	710	48.0	551	37.2	130	8.8	90	6.0	1481	1523
1975-76	801	44.4	700	38.8	132	7.3	171	9.5	1804	1896
1976-77	874	44.1	740	37.4	179	9.0	189	9.5	1982	2050

Source: Directorate of Agriculture, M.P. Agricultural Statistics, 1977.

It took about 15 years to add 95 thousand hectares to the irrigated area from 1950-51 to 1965-66, but in the next 11 years another 1,050 thousand hectares have been brought under irrigation. By the end of the Third Plan, several irrigation projects had started yielding irrigation. By the end of June, 1977, six major, 77 medium and 3,385 minor projects were completed. Consequently, irrigation potentials only from government sources increased from 158 thousand hectares to 543.2 thousand hectares of major, from 195 thousand hectares to 472.4 thousand hectares of medium and from 111 thousand hectare to 510.3 thousand hectares of minor projects, from pre-Plan period to 1976-77. Of the total created potential of 1,525.9 thousand hectares only 73.5 per cent (1,121.6 thousand hectares) could be utilised for irrigation in 1976-77.

There is a wide range of variation in the growth rate of irrigated area during 1950-77. The total irrigated area of the State increased by 126.4 per cent, while the mean of the district rates (458.5) is very high. The very high standard deviation (574) of these rates also proves very high variability. The growth rate varies from –62.0 per cent in Bastar and –19.8 per cent in Rajgarh district to 3,014.3 per cent in Rewa district (Fig. 12.6). Very high growth occurred where the initial proportion of irrigated area was extremely low. In only 12 out of 43 districts, growth rates are higher than the mean (48.5%) and eight of them (Rewa, Shahdol, Sidhi, Satna, Narsimhapur, Hoshangabad, Vidisha and Rajsen) had less than one per cent, on Dewas) between one and two per cent, and only three (Indore, Datia, Morena) had more of their cropped area under irrigation, 1950-51. Rewa recording highest increase had only 0.2 per cent of the cropped area under irrigation. The coefficient of correlation between initial proportion and rate of growth is negative and moderately significant ($r = -0.41$). The phenomenal increase in least irrigated districts is the result of a very small denominator. Even at present 10 such districts have a low to very low proportion of irrigated area. Out of 17 districts recording lower, medium growth (between mean and above 200%) four (Ratlam, Khandwa, Shajapur and Betul) also had lower medium proportion and 12 (Mandla, Surguja, Jhabua, Sagar, Damoh, Sehore, Chhindwara, Khargone, Jabalpur and Dhar) had a low and very low proportion of irrigated area in 1950-51. And thus, the quantum of denominator has been the major determinator of the growth rate of these districts also. Bhind is an exception which, in spite of a high proportion of irrigated area (10.2%) in 1950-51, recorded medium growth (285.9%).

The districts with low and very low rates of growth also present inverse relationship. There are 12 such districts, 10 of them had a high and very high proportion of area under irrigation. It increased at a very slow rate (below 50%) in four districts (Bilaspur, Durg, Seoni and Balaghat) and at a slow rate in others (Raipur, Mandsaur, Shivpuri, Gwalior, Tikamgarh and Chhatarpur). Guna and Rajgarh are two districts which, despite of an initially low proportion (2.6 and 3.9 per cent respectively), increased sluggishly. In Bastar and Rajgarh, both tribal districts in the eastern part of the State, absolute as well as percentage irrigated area declined drastically. It can well be remarked that districts with an initially poor irrigational base are still poor, despite a high and very high percentage of growth. The problem is more acute where growth is also slower with a poor base. These districts require special attention in planning of agricultural development.

Means of Irrigation

Irrigation is done with the help of canals or by lifting sub-soil water. The physical conditions have caused spatial variation in the utilisation of surface and sub-surface water resources. Canal is the major source providing water for 44.6 per cent of the irrigated area, but it is large-scale adventure and is out of the reach of individuals. Further, it cannot reach every point. It is almost single significant source of irrigation in the Chhatisgarh (Raipur, Bilaspur, Durg, Rajnandgaon and Balaghat) and in the Chambal valley (Morena, Bhind and Gwalior).

Intensity of canal irrigation is above 70 per cent of irrigated area in these areas reaching as high as 85.7 per cent in Bhind district, except in Balaghat where it is 63.4 per cent. Raisen with only 4.2 per cent of cultivated land under irrigation comes in this class. It can safely be inferred that irrigation is somewhat significant only in those areas where facilities of canal irrigation could be created. On the north-western margin of the canal-dominated area of Chhatisgarh, more than one-third but less than half of the irrigated area is provided water by canals.

Canal irrigation in the Chhatisgarh basin is as old as the first decade of the present century. At present a number of reservoirs have been constructed in this part; significant among them are Mahanadi canal, Maramsilli, Tandula, Gondli, Maroda, Kumhari, Kurung and Hasdo. In Balaghat, the Wainganga canal system provides water for more than 4,000 hectares. Other sources of canals are Sarathi, Murum nalla, Gangulpara and Kharadi. The canals constructed under the joint project of the Chambal river irrigated most of the area in its valley in the State. Next to canal is well irrigation covering 36.8 per cent of the irrigated area. Introduction of tube-well irrigation has raised the proportion of irrigated area very rapidly. In the districts of the Malwa plateau, the Bundelkhand upland and the Narmada valley, well is the major source of irrigation. In 25 districts well is the first-ranking source. It is worth mentioning that only a dew districts with a higher proportion of irrigated area than mean depend on sub-soil water resource lifted by several means. These are Tikamgarh, Chhatarpur, Shivpuri, Datia, Indore, Ratlam and Mandsaur. Along with 790,000 dug wells used for irrigation purposes, 5,359 tube-wells have been commissioned up to 1976-77. Jabalpur, Narsimhapur, Raisen, Morena, Bhind, Indore, Dhar, Dewas and Hoshangabad have significant well irrigation.

Though there are still traditional methods of water lifting in remote rural areas, diesel and electric pumping sets have penetrated into the interior areas. There were 146,997 electric pumps and 112,716 diesel pump-sets in operation in 1976-77, along with 125,741 Persian wheels. Though there are only 19.5 per cent of villages electrified, electric pumps outnumber diesel pumps. On an average, there are seven electric pumps and 5.4 diesel pumps per thousand hectares of cropped area. All districts with higher densities than means are in the western and north-western part of the State.

Wells have been the traditional source of providing irrigational water. Nevertheless the net irrigated area by them increased from 249 thousand hectares in 1950-51 to 740 thousand hectares in 1976-77, recording a three-fold increase. Their proportion in the irrigated area also increased from 28 per cent to 37.4 per cent during this period, providing water to an additional 9.4 per cent of irrigated area. This addition is more than double of the increase of canal irrigation (4.3% only). Only 12 districts have recorded a net increase in intensity of irrigation by wells: five of them are in the Chhatisgarh area (Durg, Rajnandgaon, Bastar, Bilaspur, Raigarh), two in the eastern Satpura (Seoni and Balaghat) and two in the Madhya Bharat Plateau (Gwalior and Shivpuri). Tikamgarh, Jabalpur and Sidhi are the other districts.

Tank irrigation, which had been as significant as well irrigation in the pre-Plan period, declined absolutely and relatively. At present it provides water for only 9.0 per cent of net irrigated area. It is significant only in the Chhatisgarh region, highest intensity being 48.3 per cent in Bastar. Sagar is the only district outside this area with about one-third of its irrigated area by tanks. Seoni is also significant.

There are more than 55 thousand tanks of various sizes utilised for irrigation in the State out of which 42 thousand are in Raipur and Bilaspur divisions, and more than seven thousand only in Balaghat district. Predominance of tanks in this area may be ascribed to several physical facts such as undulating terrain with inpervious rock-base and high rainfall. Water is utilised in the rainy season itself for rice, the porosity of soils does not prove hazardous. It is to be noted that fishing is also done in these tanks.

Other sources such as nallah anicuts, etc., irrigated 189 thousand hectares (9.5%) in 1976-77. It is found that the least irrigated districts invariably have a higher proportion than the average, and all districts with a higher proportion irrigated by these sources have a lower percentage of irrigated area than mean.

Irrigation and Agricultural Development

The immediate effect of the creation of irrigation facilities may be manifested in the expansion of the net area sown by annexing cultivable wasteland into it, and by intensifying the cropping pattern. The total net area sown increased from 139.7 lakh hectares in 1950-51 to 155.2 lakh hectares in 1556-57 and 185.3 lakh hectares in 1976-77. At the same time the proportion of cultivable wastelands declined from 9.6 per cent in 1956-57 to 4.7 per cent in 1976-77. There seems to be least effect on the intensification of cropping. Though its absolute area increased by about 10 per cent, its relative share declined from 13.61 per cent to 12.53 per cent of net area sown during 1956-57 and 1976-77. However, there is a very clear imprint of growth of irrigation on the increasing productivity of certain crops such as wheat, gram and sugarcane which are the major recipient of additional irrigation facilities. Expansion of area under horticultural crops is also facilitated by the development of irrigation. Further, use of fertilisers and improved seeds also seems to be associated with the irrigation.

The proportion of net area sown has increased from 32.3 per cent of total area in 1950-51 to 41.8 per cent in 1976-77. Thus 45.65 lakh hectares of land were added to net sown area while only 9.91 lakh additional area was provided with irrigation water during the same period (1950-77) which is less than one-fourth of the additional net area sown. This is suggestive that there is no significant role of irrigation in the growth of net sown area. The comparison of spatial patterns of changes irrigated area (Fig. 12.6) and net area sown (Fig. 12.4) reveals this fact clearly. Two districts recording reduction of irrigated area have presented on upward trend in net area sown, one of them, Bastar, had recorded highest growth of 224.8 per cent in the State. Four districts (Bilaspur, Durg, Seoni and Balaghat) with a very low growth rate of irrigated area have increased their net area sown very slightly, but there are a several districts presenting disproportionate changes. Guna, Chhatarpur, Shivpuri, Tikamgarh and Gwalior with a low increase in irrigated area have high and very high growth in net area sown. Similarly, Rewa, Satna, Narasimhapur, Hoshangabad, Morena, Indore, Ujjain and Datia also do not have any commensurable increase in net area sown. The coefficient of correlation is insignificant. It can, thus, be concluded that expansion in net area sown is due to other factors facilitating and necessitating expansion of cropped area rather than to increasing irrigation facilities.

Impact of Irrigation on Cropping Pattern

Farmer always wants to maximise return from farming and, therefore, he uses such inputs as irrigation and fertilisers for those crops which are capable of giving commensurable returns. From this simple logic it can be deduced that the major irrigated crops should be superior crops, and that the irrigated area of only those crops increased rapidly which could rise their productivity proportionately.

Out of the total irrigated are (2,050.3 thousand hectares) more than three-fourths are under only two crops — wheat and rice. Other beneficiaries are horticultural crops (6.57%), gram and sugarcane. It is noteworthy that the proportion of irrigated area under sugarcane, fruits, vegetables and spices is many times higher than their proportions in total cropped area, as is clear from Table 12.2.

TABLE 12.2: Irrigated crops in M.P., 1976-77

Crop	Proportion in cropped area	Total irrigated area	% irrigated area under crops	% change in irrigated area 1950-77
Wheat	15.1	40.0	26.1	589.3
Rice	22.2	35.6	15.8	32.1
Gram	9.7	5.8	5.9	302.2
Sugarcane	0.4	3.6	96.6	124.2
Fruits-Vegetables	0.8	3.5	49.4	—
Spices	0.6	3.1	50.0	—
Barley	0.8	2.8	—	8.8
All crops	100.0	100.0	9.8	126.4

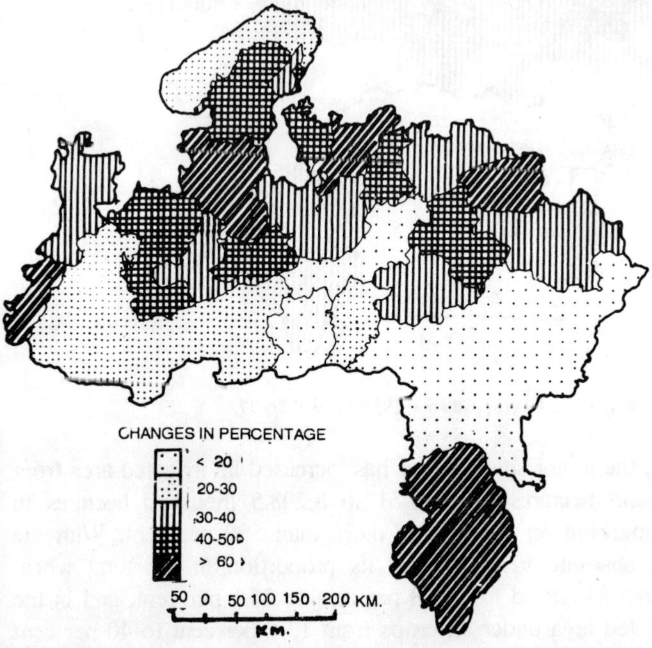

FIG. 12.4: Changes in net sown area, 1950-51 to 1976-77

There is marked regionalism in irrigated crops (Fig. 12.8). In the irrigated tract of the Chhatisgarh basin, rice is the only major irrigated crop. In other parts, though wheat is first-ranking irrigated crop, other crops are also significant, such as barley and gram in the Bundelkhand upland and Madhya Bharat Plateau, horticultural crops, sugarcane and cotton in western Malwa and in the Narmada valley. Two crops, opium in Mandsaur and rape and mustard in Morena, are unirrigated crops in the State. This proves that those crops are significant in the role of irrigated ones which have a high calorific as well as market value.

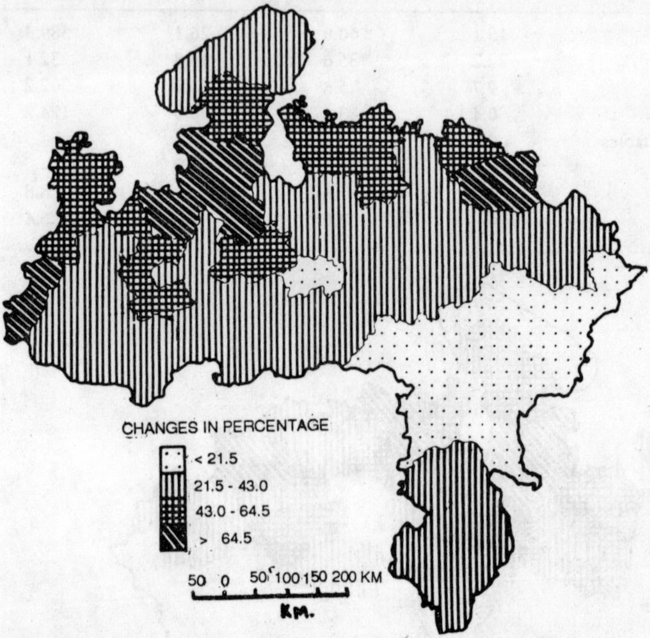

FIG. 12.5: Changes in total cropped area, 1950-51 to 1976-77.

Wheat, the major share-holder has increased its irrigated area from 119 thousand hectares in 1950-51 to 8,203.5 thousand hectares in 1976-77, marking an increase of more than 589 per cent. With the increasing absolute irrigated area its proportion in the total wheat cropped area increased from 5.8 per cent to 26.1 per cent, and in the gross irrigated area under all crops from 13.1 per cent to 40 per cent during the same period. As a consequence of rapid expansion of the irrigated area, its production increased by 117.9 per cent while the total area under it by only 52.1 per cent. Its yield per hectare was only 512 kgs in 1950-51 which reached 766 kgs in 1976-77. These two results are seen in all districts. All districts recorded increase in the irrigated area under wheat and in 32 districts yield per hectare also increased. In other 13 districts it declined with increasing irrigated area because of the fact that rainfall in these districts was remarkably lower than the average which necessitated expansion of irrigation also. Even in these districts, yield rates were much higher in the previous year, i.e. 1975-76. It has been found that rapid growth occurred after the Second Five Year Plan, after the launching of the Green Revolution Movement in the State.

FIG. 12.6: Madhya Pradesh: Changes in irrigated area during 1950-76.

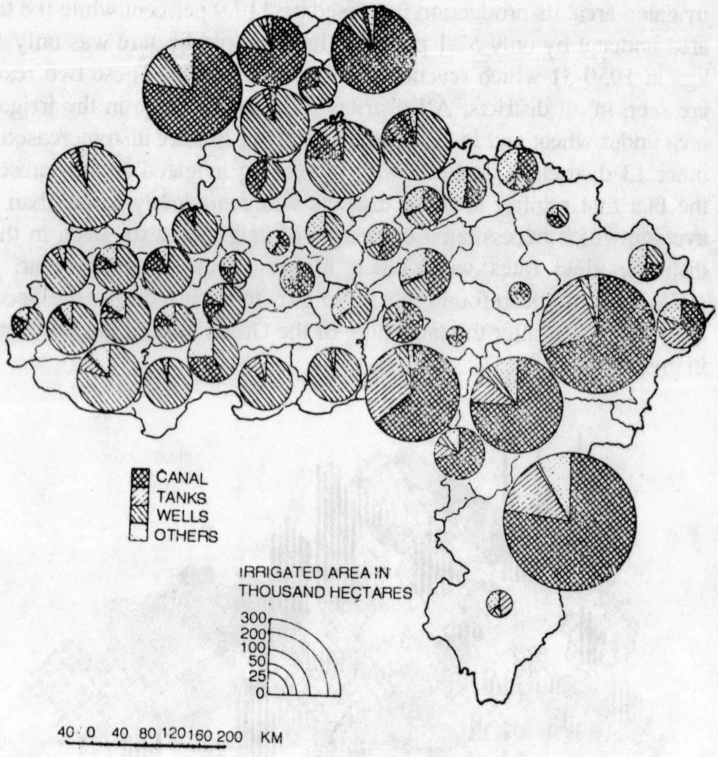

FIG. 12.7: Madhya Pradesh: means of irrigation, 1976-77.

Rice is next to wheat sharing more than one-third of the total irrigated area. As a result to very sluggish growth of 32 per cent from 552.6 thousand hectares in 1950-51 to 730 thousand hectares in 1976-77, the share of rice in gross irrigated area declined drastically from 61.0 per cent to 35.6 per cent during this period. Despite this, production of rice increased by 104 per cent from 1,369 thousand tonnes to 2,797 thousand tonnes and the yield rate by 78.7 per cent from 357 to 638 kgs/hect., leaving wheat far behind. There are several districts (18) which witnessed reduction in actual rice irrigated area but recorded a remarkable rise in production as well as in yield. It seems rational that development of this crop grown in the rainy season depends upon other inputs such as high-yielding variety of seeds, etc., than on irrigation. However, in scarcity years irrigation plays a vital role. At the same time, rapid expansion of rice-cropped area in the north-western part of the State has been facilitated by the irrigation.

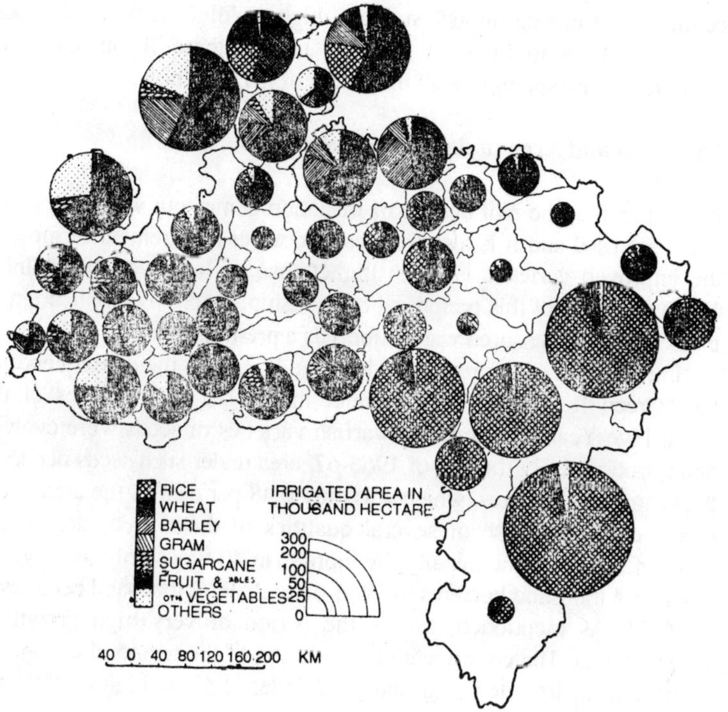

FIG. 12.8: Madhya Pradesh: irrigated crops, 1976-77.

Gram, a hardly crop, also recorded very high (302%) increase in its irrigated area, from 29.4 thousand hectares in 1950-51 to 118.2 thousand hectares in 1976-77. Only 5.8 per cent of gram cropped area is under irrigation and hence changes in its yield rate cannot to ascribed to it. Sugarcane, condiment spices and fruit-vegetables are highly dependent on irrigation. Sugarcane is an almost entirely irrigated crop with more than 96 per cent of its cropped area under irrigation, recording increase very near to the average of all crops (124.2%). There is a very small change, of course upward, in the yield rate (9%).

Irrigation and Adoption of Yield-Raising Inputs

The preceding analysis makes it clear that irrigation has played a vital role in enhancing the yield rates of the crops rather than in colonisation in the State. At the same time, it has also influenced the adaptability of certain yield-raising inputs such as high-yielding variety of seeds, fertilisers. It is to be seen how far use of these inputs co-varies temporarily and spatially with the extent of irrigated area.

Irrigation and Area under HYV of Seeds

It is a fact that the real benefit of irrigation comes out when improved seeds are used and it is also true that the water requirement of most of the improved varieties is much higher than of local varieties (Sinha, 1975). Because of this reciprocal relationship, before the application of improved seeds, ensured water supply is a pre-requisite.

Introduction of improved seeds in the State was the consequence of the "Green Revolution" programme. In the last two decades of the Third Five Year Plan several dwarfish varieties of seeds were evolved and introduced. By the end of 1966-67, area under such seeds occupied 57.4 thousand hectares, which was only 0.48 per cent of the area under cereal crops. Because of several qualities of these seeds area under them increased phenomenally, by more than 40 times only in 10 years, from 57.4 thousand hectares in 1966-67 to 2,597.2 thousand hectares in 1976-77. As mentioned, this is the period of very high growth in irrigated area. The coefficient of correlation of the temporal changes in proportion to irrigated area and area under HYV is high (r=0.9) and significant.

At present about 21 per cent of the area under cereals has been cropped with HYV of seeds. This area is certainly much higher than the proportion of irrigated area (13.0%) under them. Despite this lacuna, there is a close comparability in the spatial distribution of these two phenomena. Figure 12.9 shows the concentration of HYV of seeds in four tracts, viz. the Chhatisgarh basin — Raipur, Durg, Rajnandgaon and Balaghat; the Chambal sub-humid area — Bhind, Morena, Gwalior, Datia and also Tikamgarh; the Narmada valley — Narsimhapur, Hoshangabad, Khandwa and Khargone and the western Malwa — Mandsaur, Ratlam, Shajapur, Indore, Dewas and Bhopal districts. Most of these districts are on the upper side in respect of irrigated area. There are a few exceptions. Bilaspur and Shivpuri districts with a higher proportion of irrigated area have comparatively lower area under HYV, while inverse is the case with Rajnandgaon, Narsimhapur, Khandwa, Khargone and Bhopal. Other districts with low irrigation facilities also have little area under HYV of seeds. As a result of this concurrence, the coefficient of correlation is significantly high (r = + 0.70). It seems that irrigation has a decisive influence on the use of HYV of seeds.

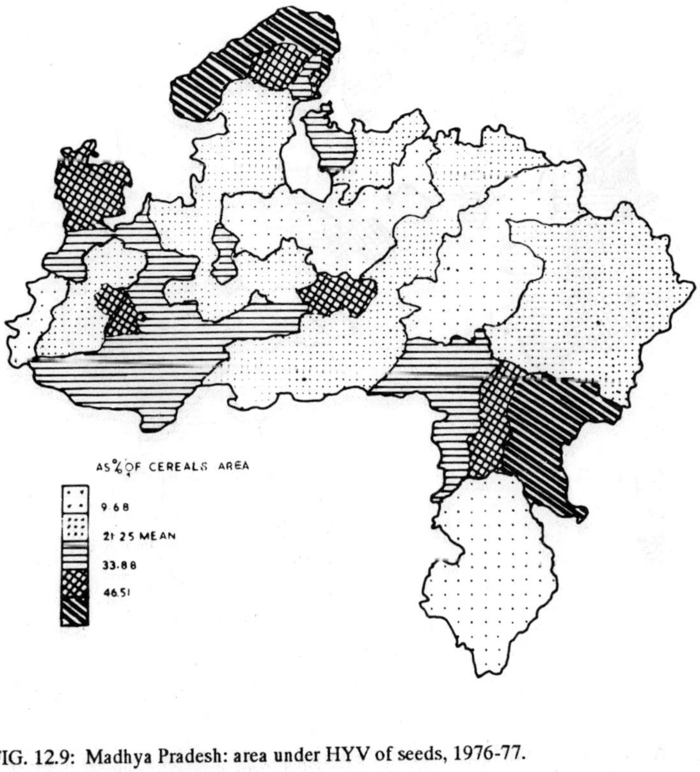

FIG. 12.9: Madhya Pradesh: area under HYV of seeds, 1976-77.

Irrigation and Use of Fertilisers

Fertilisers are not popular in this State. On average, 6.56 kgs of chemical fertilisers are applied per hectare of cropped area in this State which is a little more than one-third of the average of the country. Poverty of farmers, lack of ensured timely water supply and the subsistence nature of cropping may be the reasons of this State of affairs. Nevertheless, distribution of fertilisers has recorded an increase of more than 53 times, from 2.96 thousand tonnes in 1956-57 to 157.94 thousand tonnes in 1976-77. The annual average of use of fertilisers per hectare increased from 0.17 kgs to 6.56 kgs during the period. This growth is very similar to that of irrigated area.

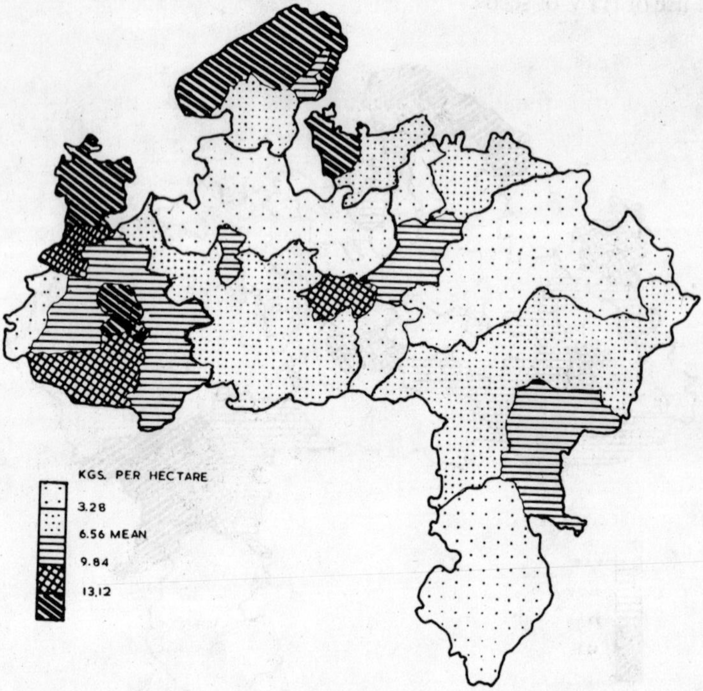

FIG. 12.10: Madhya Pradesh·

There is a wide variation in the spatial pattern of use of fertilisers. It ranges between 0.51 kg per hectare in Bastar and 20.34 kgs per hectare in Gwalior in 1976-77. There are two areas of concentration: first, the Chambal sub-humid area (Gwalior, Bhind, Morena, Datia and up to Tikamgarh), and the western part of the State comprising Mandsaur, Ratlam, Ujjain, Indore, Dewas, Khargone and Khandwa. Jabalpur and Narasimhapur from another pocket of high concentration. Outside of these areas are Bhopal and Raipur. If the map (Fig. 12.10) showing use of fertilisers is compared with the map of irrigated area, it is evident that, except in the Chambal area, in most of other districts the proportion of irrigated area, though very close to the average, is lower medium. They are getting lower returns than the irrigated districts, conversely, highly irrigated districts of Chhatisgarh have a lower average than of the State. It evidently shows that fertilisers are less popular in the eastern part of the State dominated by Kharif crops with rice as an important crop than in the western part of dominantly Rabi crops. Further, it is noteworthy that these western districts are significant in cotton cultivation. However, the correlation coefficient between irrigated area and use of fertilisers is moderately significant ($r = 0.56$).

Conclusion

The State of Madhya Pradesh is agriculturally backward. Like other developmental measures, development of irrigation is a localised phenomena. There seems to be no direct relationship between the amount of rainfall and intensity of irrigation. The percentage of irrigated area increased rapidly where its intensity was low and very low in the initial years, i.e. in 1950-51. The expansion of net area sown and reduction of cultivable wastes cannot be ascribed safely to the development of irrigation. However, irrigation has played a vital role in raising the yield rates of certain crops, such as rice and wheat and horticultural crops. This is made possible through the adoption of certain yield-raising inputs, which are dependent, to a great extent, on irrigation. Use of the high-yielding variety of seeds and chemical fertilisers varies directly with the intensity of irrigated area and have a significantly high positive correlation. The districts adopting these three measures, viz. irrigation, HYV of seeds and fertilisers also have a high agricultural productivity per unit of cropped area. It may be summed up that the development of agriculture in the State is tagged with the development of irrigation. Only the surface water resource is partially used for the purpose, leaving reserves of the underground water resource untapped. Therefore, use of the water resource may be planned accordingly.

13

HIGH TECHNOLOGY IRRIGATION IN AUSTRALIA

TOM McKNIGHT

HIGH TECHNOLOGY irrigation is a well-established phenomenon in Australian agriculture. This should come as no surprise to even a casual observer because there is a lengthy history of inventing/adopting/diffusing agricultural innovations in Australia and because irrigation has long been so significant in the agrarian scene.

From the earliest years of commercial farming there have been imaginative attempts to accommodate the exigencies of the Australian environment to the needs of the man on the land. Individual farmers and engineers, often stimulated by government-sponsored contests and prizes, have come up with a remarkable variety of devices and machines to ease their tasks or make their production more efficient. Beginning with the stripper wheat harvester in the 1840s, such inventions as power shears, the stump jump plough, the header harvester, and the rotary hoe have continued to embellish the Australian reputation for imaginative innovations in farm machinery.

Irrigation was not an early development in Australian agriculture, but once it was initiated, its spread was rapid and its significance to the farm economy became notable. Several small irrigation schemes were started in Victoria in the 1870s, but large scale activities date from the next decade, led by the Chaffey Brothers in the riverlands of the middle sector of the Murray.

Irrigation has, of course, been a basic farm operation for centuries in many parts of the world. Its major attribute is that it greatly enhances crop yields. One of its principal disadvantages is that it normally requires a considerable input of arduous labour to prepare the field and/or introduce water to the crop. Thus one of the very appealing features of most high technology irrigation is that the labour requirement is notably diminished.

Irrigation technology has experienced dynamic development for the past four decades. The modern irrigator—beset by scarcer water supplies, more costly energy, and less reliable labour—generally must choose between the simplicity of gravity-flow surface irrigation or the proficiency of automated sprinkler equipment. During the past quarter century the trend clearly has been toward the latter option—complex automatic machinery that sprinkles the crop from above.

In the Australian context, high technology irrigation comes in three modes:

(1) Laser-levelling of the land for expanded/enhanced flood or furrow irrigation. The actual irrigation method is not high tech, but the method of preparing the field employs a very sophisticated procedure. Laser-levelling was first utilised on a large scale in the American state of Arizona about 15 years ago. It is expanding considerably in Australia, but apparently there are no quantitative data available to indicate the actual extent of its use.

(2) "Drip" or "trickle" irrigation was developed prominently in Israel several decades ago. It brings water to the crop by means of low-volume conduits (usually plastic hoses) that have many small orifices through which the water is emitted either continuously or at regular intervals. In recent years various sophisticated machines have been added to some drip systems to maintain more efficiently a balance between the moisture needs of the crop and the water intake rate of the soil. These high technology drip systems are usually referred to as "micro irrigation". In Australia, as elsewhere, micro irrigation is mostly employed in orchards and vineyards, although there is increasing use in vegetable production. Today there are about 30,000 hectares of drip irrigation in Australia, but there are no available statistics to indicate what proportion of that total is high technology micro irrigation.

(3) Complex automatic sprinkler machines comprise the principal form of high technology irrigation in Australia and constitute the major concern of this paper.

Sprinkler irrigation made its first appearance about the beginning of the present century. Usually it involved permanently installed overhead spray lines on high-value crops until after 1939-45 war, when lightweight aluminium pipe became available at reasonable prices. Sprinkler technology then began to develop and diversify, especially in the U.S.A., based particularly on the use of rotating sprinkler heads, quick couplers for lengths of pipe, and impulse sprinklers. All sorts of imaginative derivations of sprinkler systems were produced in the late 1940s and early 1950s.

The most far-reaching systemic development was a truly automated continuous-move system, which was invented in 1949 by a tenant wheat farmer in Colorado. His device was dubbed "centre pivot", a name that has persisted. Centre pivots were first marketed in the early 1950s in Nebraska, the state which is still the hub of both manufacture and use of centre pivots. These machines began to be adopted widely in the 1960s, primarily in the Great Plains. They became a national phenomenon in the 1970s, expanding since then by about 1000 per cent in the United States, where they now water some 16 per cent of all irrigated land, as well as more than 10 per cent of all irrigated hectarage in Canada. Centre pivot continues to be the fastest growing segment of the irrigation industry in both countries.

Centre Pivot Technology

The principle of centre pivot technology is complicated, and the machines themselves are, perhaps, the most sophisticated that farmers are ever called upon to operate. A centre pivot machine consists of a series of metal pipes connected end-to-end to form a lateral that is supported by evenly-spaced wheeled towers. The lateral is anchored at one end, and water is fed into the machine at that point. The machine rotates around the pivot point, dispensing water through openings that are dispersed along the lateral. The lateral is suspended from one to five metres above the ground so that it will clear field crops as it rotates. The towers move independently, and perhaps, the most noteworthy achievement of centre pivot technology is the sophisticated alignment mechanism that maintains a straight-line advance as the machine rolls along its predesigned route. A centre pivot system is a truly automatic device that requires minimal labour, while distributing water in equal amounts over the entire circular area.

The benefits of centre pivot are notable, and particularly include the following:

(1) There is an enormous reduction in labour requirements. Once

the machine has been installed, it is virtually a push-button operation. Thus, a single worker can manage the irrigation of literally hundreds of hectares without assistance.

(2) Water management is extremely efficient. The machine can be programmed precisely for volume of flow, size and spacing of orifices, speed of movement and other parameters that allow careful satisfaction of crop water needs over the entire wetted area. No other form of irrigation produces such efficient use of water.

(3) The machines can function on undulating or sloping land, so that land preparation efforts and costs can be greatly reduced. Flat land is not necessary for centre pivot operation.

(4) Sandy soil often cannot be irrigated by conventional methods because its high porosity absorbs too much water at the "head" of the field, and not enough moisture reaches the "foot". However, a continuously-moving centre pivot machine brings water to all parts of the field equally.

Despite these undeniable advantages, centres pivots have not been an unblemished success. There are three major problems:

(1) The initial investment is high. The average centre pivot machine that will water 50 to 60 hectares costs about $115,000 in Australia.

(2) Centre pivots water in a circular configuration. As most fields are rectangular in shape, this leaves the corners unwatered. In a square field, about 17 per cent of the area is left dry. Various devices can be added to the machine to bring water into the corners, but these are costly.

(3) The outer portion of the lateral moves faster than the inner portion, so water must be dispensed in greater volume there in order to provide uniformity of application over the entire circle. This may result in excessive runoff, rutting in the wheel tracks, and even bogging of the entire machine. This is a particular problem where soils are "heavy" (i.e., have a high clay content).

Evolution of Linear Move

Although centre pivot technology was widely adopted and generally successful (spectacularly successful in some areas), there were also some notable misadventures. Its greatest inadequacy surfaced in the Central Valley of California, which is America's leading irrigation area. Much of the Central Valley has heavy clay soils, and their slow water intake rate simply could not absorb the high volume of application in the outer portion of the circle, resulting in conspicuous failures.

Spurred by this nagging problem, some of the major manufacturers developed an even more complex machine, called "linear move" or "lateral move". It waters in a rectangular configuration, with the entire machine moving at the same speed over the field, thus obviating two of the three difficulties described above. Linear move is based largely on centre pivot technology; the greatest new engineering demand being the problem of introducing water into an unanchored, continuously-moving machine. Three different techniques were devised to meet this challenge, each at considerable cost.

The typical linear move machine is considerably larger than a centre pivot, and the area watered is significantly greater (Fig. 13.1). Per hectare costs, however, average about the same.

The prototype linear move machine was put to work in the Central Valley in 1977, and was well received by local agricultural operators. At the present writing, one decade later, there are about 175 linear move machines functioning in California (most of those in the Central Valley), and an approximately equal number scattered elsewhere in the United States.

Developments in Australia

It turns out that Australians were early on the centre pivot scene, although their influence does not seem to have been great. A part-time farmer with a small property near Sydney developed a pivot-centred irrigation machine in 1953; an invention that was totally independent of any American activity. It used a water-driven piston and a rachet wheel for continuous move, watering about three acres (Fig. 13.2). Wanting a larger unit, he continued tinkering for another three years and finally produced a substantial centre pivot that had self-supporting pipe with stages underneath.

He formed a company (Grasslands) which developed an association with a large agricultural pump manufacturer in California. Over the next dozen years about 40 centre pivot systems were produced and sold by Grasslands, including a few exported to the United States. Some of

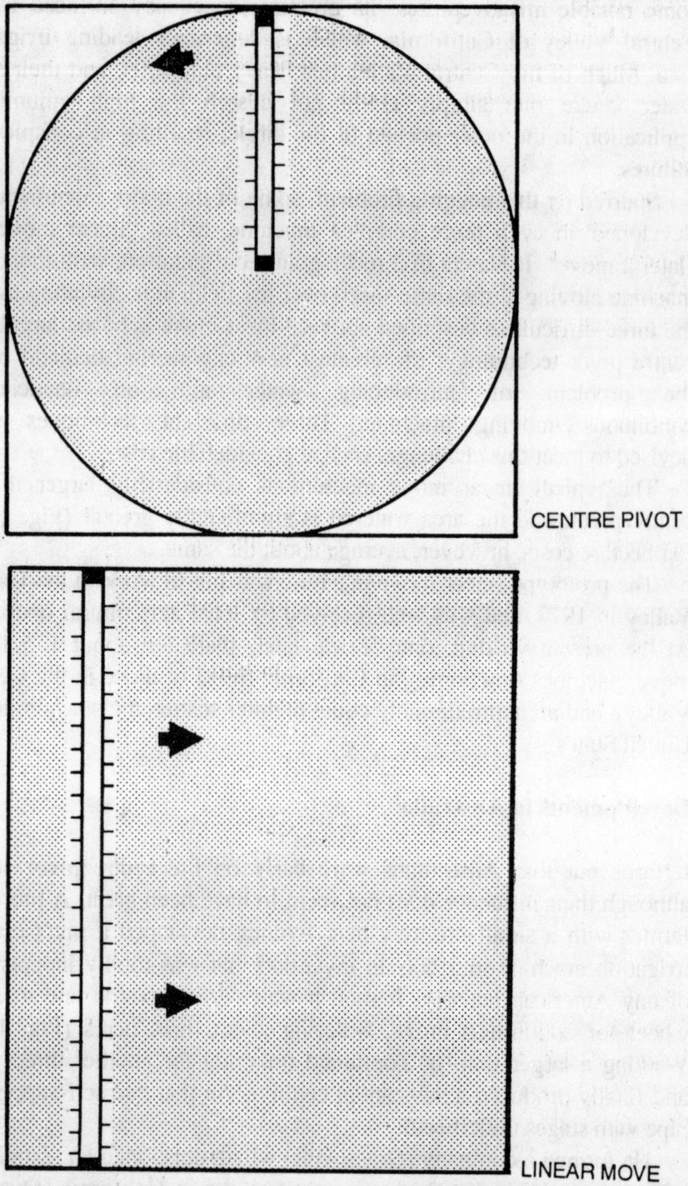

FIG. 13.1: Diagrammatic representation of the configuration of centre pivot and linear move machines. The shaded portion is the area actually watered by each machine

FIG. 13.2: Jack Allwood's original centre pivot machine, near Sydney, 1953. (Photo courtesy Jack Allwood)

these machines were fairly primitive and rather ineffective, but there is at least one instance of two Grasslands pivots that were installed in 1958 and continued in use until 1984, when they were replaced by new machines. Systems produced by Grasslands were marketed fairly widely, with most sales being in the Moree area of New South Wales and the Border-town district of South Australia. Grasslands went out of the centre pivot business about 1970.

Centre pivots from America began to be imported to Australia in the early 1970s, primarily to north-central New South Wales (the Moree-Collarenebri area). There was a slow but continuous introduction of American pivots through the late 1970s. Until about 1977 probably 90 per cent of all Australian pivots were located in New South Wales, particularly in the north-central district. The first centre pivot in the MIA was not installed until about 1975. Beginning in the late 1970s there was a significant upsurge of installations in South Australia, especially in the riverlands of the Murray drainage, and to a lesser extent in other parts of New South Wales and in Queensland (Fig. 13.3).

The first linear move machine was imported and installed near Moree in 1979; another was placed near Warren in the same year. This was quite early on the scene, as the California prototype machine had only been in existence for less than two years (Fig. 13.4).

Basis of This Study

This paper is based primarily on research carried out in 1985. As there is almost no literature on automated sprinkler irrigation in Australia, the methodology mostly consisted of personal interviews, telephone interviews and postal questionnaires. Various irrigation officials and farm machinery manufacturers and distributors were interviewed, but the bulk of the data was obtained through postal questionnaires (along with some follow-up interviews) with farmers who use centre pivot or linear move machines.

About 40 interviews were carried out, and 105 questionnaires were circulated. There was a gratifying 57 per cent response to the questionnaire survey. It was not, however, a balanced response. More than half of the respondents were South Australians and only 15 per cent of the replies came from New South Wales, which is the leading centre pivot state.

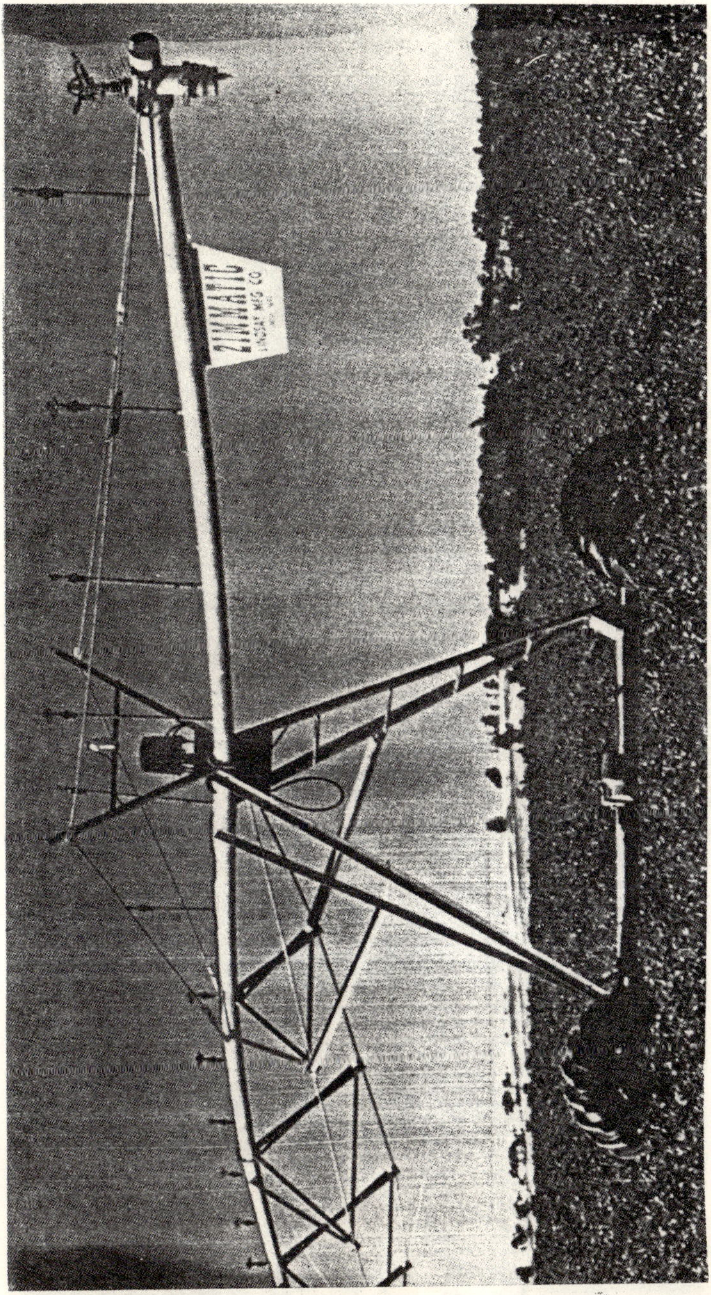

FIG. 13.3: A typical centre pivot machine on potatoes, near Leeton, NSW, 1984.

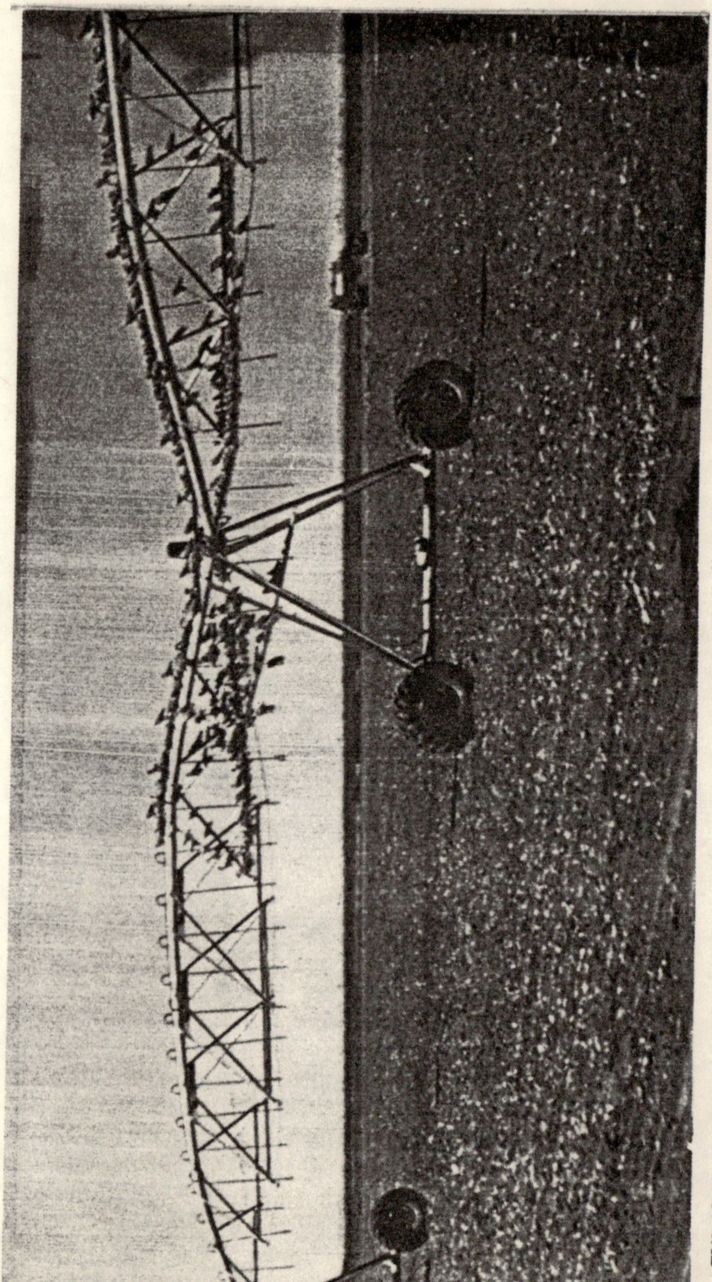

FIG. 13.4: A linear move machine near Hay, NSW, 1984. The laterals and struts provide good perches for a mob of galahs.

The production/distribution industry in Australia is relatively small. There are three centre pivot manufacturers—located at Murray Bridge (S.A.), Toowoomba (Qld.), and Bundaberg (Qld.)—and at least one of them also produces linear move machines. In addition, there are two other major distributors (for U.S. machines), and about eight smaller distributors.

Most of the data presented and discussed below are derived from this survey. As the survey is both incomplete and unbalanced, it cannot be considered as definitive. It does, however, represent information from a variety of sources, and is believed to be broadly accurate.

Distribution

Leading irrigation authorities estimate that in Australia there are between 350 and 400 centre pivot systems and perhaps 30 linears. This would amount to a total area of about 30,000 hectares, or two per cent of all Australian irrigated area.

My incomplete survey discovered about 330 pivot systems and more than 50 linears. I judge that this represents only about 60 per cent of the pivots actually in operation, and perhaps 90 per cent of the linears. Thus my best estimate is that there are now functioning in Australia at least 500 centre pivots and between 55 and 60 linears. This would comprise about 50,000 hectares, or more than three per cent of total Australian irrigated area.

It appears that about 40 per cent of all pivots are located in New South Wales (Fig. 13.5). They are widely distributed in the central part of the state, especially in the Moree, Dubbo-Narromine, and Jerilderie-Deniliquin areas (Fig. 13.6). Centre pivots are still not very common in the MIA.

South Australia probably contains one-third of all Australian pivots, mostly in the Lower Murray district, the Bordertown area, and the vicinity of Renmark. There are smaller numbers in Queensland, especially around Emerald. Pivots are widely scattered in northern Victoria.

In Western Australia the first units were introduced in the early 1970s—two at Gingin and two on Anna Plains station near Broome. None of these proved to be very satisfactory. All other Western Australia pivots apparently were installed in the early 1980s; there is a total of about 15 systems in the state. The Northern Territory contains less than half a dozen units.

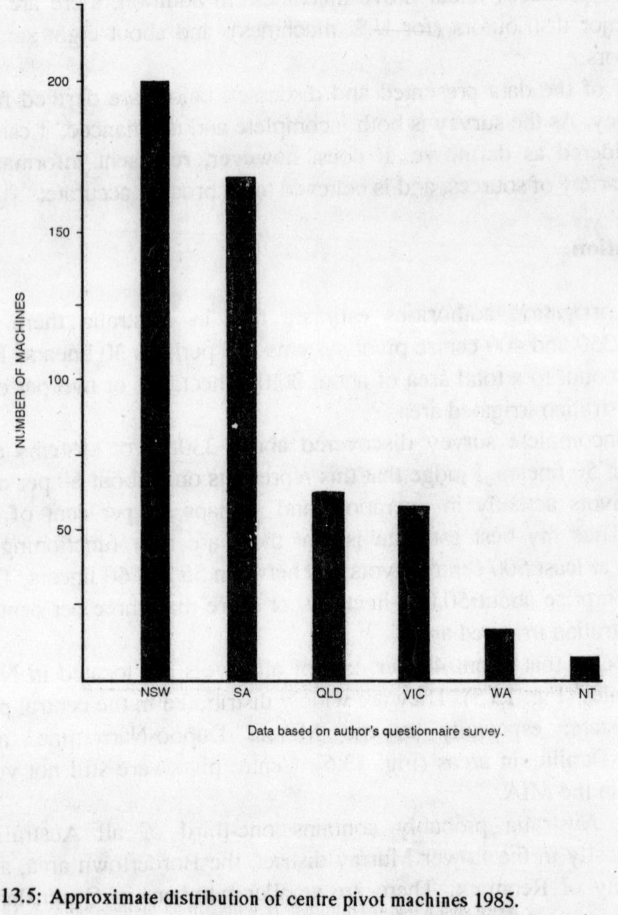

Data based on author's questionnaire survey.

FIG. 13.5: Approximate distribution of centre pivot machines 1985.

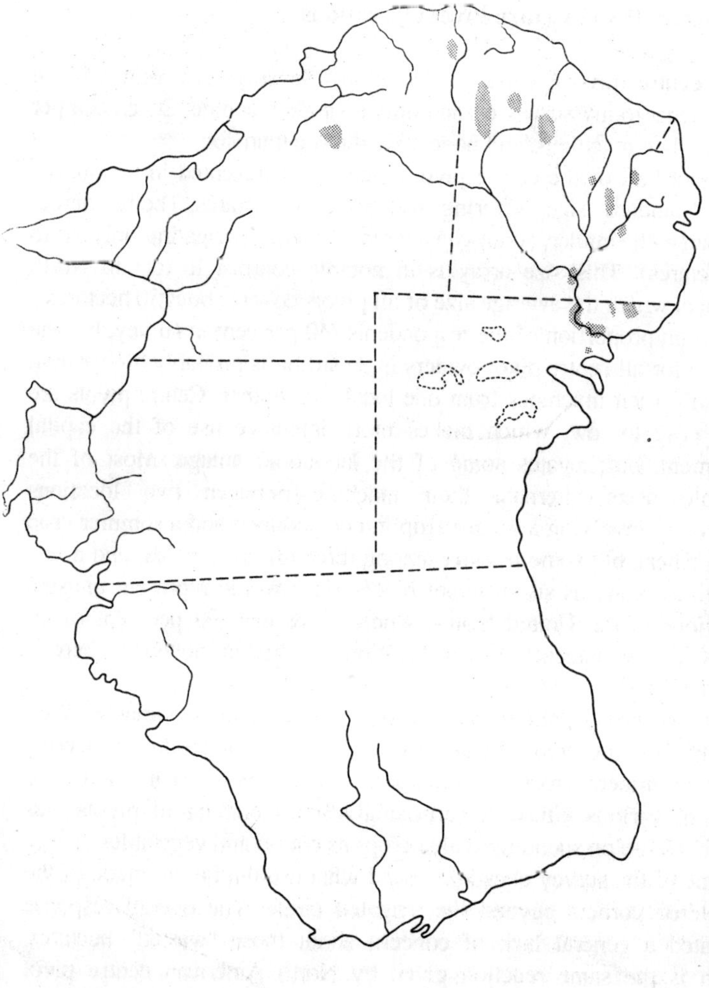

FIG. 13.6: Shading indicates the principal concentrations of centre pivots in Australia. (Based on author's questionnaire survey)

New South Wales is even more dominant in linear move distribution; at least 75 per cent of Australian linears are in that state. They are mostly located around Moree, Hillston, and Jerilderie. In other states they are widely scattered, including an installation of six on one property in Western Australia.

Characteristics of Centre Pivot Operations

Most centre pivots exist in isolation. Seventy per cent of the respondents to my survey owned only a single machine. Seventeen per cent had two machines, and none owned more than six.

About half of the centre pivot machines represented in the survey are of "standard" size, watering from 50 to 70 hectares. The remainder are nearly all smaller, some considerably smaller (irrigating only 20 to 30 hectares). This size array is in notable contrast to that in North America, where the average size of all pivots covers about 50 hectares.

A high proportion of the respondents (40 per cent in survey, but the average for all centre pivot owners in Australia is probably higher than that) shift their machines from one locale to another. Centre pivots are fairly easy to tow, which makes more intensive use of the capital equipment, but negates some of the labour advantage. Most of the multiple users alternate their machine between two locations (frequently involving a winter crop in one paddock and a summer crop in the other), but some transfer among three different fields, and a few shift to as many as six different places. This would represent unusual behaviour in the United States, where more than 90 per cent of all pivots stay permanently in one location; in Canada, however, there is more shifting.

The cropping pattern under centre pivots is quite variable. Well over half of the area irrigated by these machines produces lucerne. Other prominent crops are potatoes, wheat, other cereals, and seed crops of various kinds. A surprisingly small number of pivots was reported to be on such high-value crops as cotton and vegetables.

One of the survey questions asked what use the farmer made of the unwatered corners beyond the irrigated circle. The overall response indicated a general lack of concern about those "wasted" hectares, which is the same reaction given by North American centre pivot users. Some Australian respondents graze the corners; others carry on dryland cropping; about ten per cent attach "end guns" to the extremity of the pivot to shoot water into the corners as the machine rotates by.

In an attempt to gain some insight into the diffusion of information, users were asked where they learned about centre pivots in the first place. More than half replied that they had seen the machines in operation on other farms (mostly on nearby farms, but in some cases on a North American excursion). The only other prominent source of information consisted of advertisements or stories in rural magazines and newspapers.

The general reaction of users to their pivots was positive. Seventy-five per cent reported "good" or "very good". Only eight per cent stated "bad", whereas the remainder had a mixed response.

When queried about problems, most respondents alluded to minor mechanical difficulties, or trouble with the application rate. There was little mention of any serious rutting/erosion/bogging problems. For many farmers it is obviously quit a challenge to learn to use these sophisticated machines, and also to ascertain precise water needs of their crops and water intake capabilities of their soils.

The advantages listed by respondents were as expected—limited labour need, efficient water use, ease of operation, use on uneven or sandy land, lower energy cost, and particularly enhanced productivity.

The final question asked the user if he planned to add more centre pivots to his operation. Twenty-five per cent answered affirmatively; another 40 per cent stated they would add more pivots if they had the land and/or water to support them.

Some Characteristics of Linear Move Operations

A linear move enterprise generally represents irrigated farming on a grand scale. The typical Australian linear waters from 400 to 700 hectares, which is ten times the size of the average Australian centre pivot operation.

It is also considerably more sizable than the typical American linear move undertaking. The machines themselves are roughly the same size as American linears; about 800 to 1000 meters in length. However, the average "run" is much longer. The typical American linear will have a one-mile (1600 meters) run, whereas most Australian linears run twice that far, and some go five times that.

In a move toward more restricted size, a few Australian operators have opted for the "swing-around" linear, which uses a machine only half as long. It can be re-positioned at the end of the run to water an adjacent rectangular field. This allows for smaller areas to be watered economically.

Other characteristics of linear move operations (water source, crop choice, etc.) are much the same as centre pivot operations.

The Outlook

The future of high technology irrigation in Australia will depend largely on the strength of the market for farm commodities. When the market is depressed, there is little enthusiasm for such large capital investment as is required for a centre pivot or linear mover system.

However, even in a depressed market, some farmers are more adroit than others in adjusting to economic realities. And it is abundantly clear that a clever, or lucky, farmer can use high technology irrigation to good advantage. These machines provide the most efficient techniques for water use, and this is a critical consideration in a dry continent.

Short-term indications seem to point toward modest expansion of both centre pivot and linear move irrigation, with perhaps 50 to 100 new systems being installed annually.

REFERENCES

Tom L. McKnight, 1983: "Centre Pivot Irrigation in California," *Geographical Review*, Vol. 73, January, pp. 1-14.

———, 1979 : "Centre Pivot Irrigation: The Canadian Experience," *Canadian Geographer,* vol. 23, October, pp. 361-67.

———, 1979 : "Great Circles on the Great Plains: The Changing Geometry of American Agriculture," *Erdkunde,* Vol. 33, pp. 70-79.

———, 1986 : "The Development of Linear Move Irrigation in the United States," *Yearbook,* Association of Pacific Coast Geographers, Vol. 48, (in press).

William E. Splinter, 1976 : "Centre-Pivot Irrigation," *Scientific American,* Vol. 234, pp. 90-98.

14

CASTE STRUCTURES AND UTILISATION OF WATER RESOURCES FOR IRRIGATION

DEBASHIS DAS

AGRICULTURAL DEVELOPMENT of any region depends mainly upon the availability of water resources for irrigation along with the favourable quality of land and soil resources, other technological inputs, machineries, and institutional arrangements. In a developing country like India, irrigation plays the role of a catalyst and can assenable the necessary inputs and innovations along with it for the use in agriculture. At the same time the significance of social factors cannot be over-ruled, which influence the utilisation of water resources in the entire region. Among the social factors the socio-economic relation and functional dependence between the farmers belonging to different caste groups, deserve worth mention. Often and on it is surmised that due to unhealthy functional relations between the farmers of different socio-economic groups of interests the available water resources are not always properly used, rather a serious tussle is commenced for hankering for the water resources. According to Thorner and Thorner (1961), the "strong groups" always get the better of the weaker sections of the society and their power always remains as unchallenged. The above conflict encourages to waste water resources and to decline productivity value of crops in the region. In order to probe further into this issue, the district of Birbhum (W. Bengal) has been chosen.

Aims and Objective of Study

The main aims and objectives of the study are as follows:

(a) To find out relationship between general castes, scheduled castes and scheduled tribes population in different blocks of

Birbhum district.

(b) To identity relationship between population, belonging to different castes groups and irrigation in different blocks of Birbhum district.

(c) To find out relationship between population, belonging to different castes groups and net irrigated area, net value of crops at different sample villages of the district.

(d) To identify the relation between population, belonging to different castes groups and sourcewise irrigated areas in the district.

Methodology

The study is entirely based on primary data collected from 15 villages selected randomly for intensive household survey. From each village twenty four households are selected using purposive sampling technique, and they were thoroughly interviewed according to a well structured questionnaire.

The Study Area

Birbhum district is located at the western part of the state. It is bounded on the north by Malda district, on the south by the Ajoy river, on the east by Murshidabad district and on the west by Chotanagapur plateau of Bihar. The Ajoy river marks its southern boundary. The total area of the district is 4545 km^2 where 2095829 population reside. The density of population is 461 per km^2. The entire region is a monotonous plain with some undulations on the west. The average height of the district is 195 feet above mean sea level, and the highest part (320 feet) is located in the west, where as the lowest part (70 feet) is on the south east. This implies that the general slope of the region is from west and north to east and south east. The average slope of this district is 6 feet per mile.

The district is covered with sandy loam, sandy clay, loam, and clay soils. On the west, laterite soils predominate and older alluvium. These soils are fertile and became highly productive if irrigation and other inputs are applied to it in an appropriate measure and manner.

The region has both the surface and ground water resources. Surface water resources are utilised through canal and tank irrigation and the groundwater resources are tapped by tubewell and well irrigation. Besides, the river water is also utilised by means of lifting devices. The district has 3,90,054 acres of land under irrigation which is 58.97 per cent of the net sown area. Canal irrigates an extensive area without being facilitated by the field channels. Next important source is the tank which is found in entire part of the region, where as the former one is found only in northern central and eastern parts. Tubewell irrigation predominates in the central and eastern parts of the region and along the river valleys. Whereas the river lift irrigation schemes are operated only on the banks of the rivers. Diverse water resources are available in the district yet due to the absence of proper management practices, these water resources are not fully utilised for the development of agriculture in this district.

The district is a dwelling place of farmers, belonging to different caste groups. These includes high caste Hindu (Brahmin, Baidya, Kayestha), middle caste Hindu (Sadgope, Ugrakhatriya, Mayra etc.) scheduled castes (Bagdi, Bauri, Dome etc.), scheduled tribes (Santal, Konra etc.) and Muslims. It became evident from a sample survey that out of the total population 4.66 per cent belong to high caste Hindu, 30.20 per cent belong to middle caste Hindu groups. The scheduled castes and tribes constitute 34.33 per cent and 7.66 per cent of the total population respectively and rest 23.15 per cent of population belongs to Muslims. The high and middle caste Hindu population and some muslims are socio-economically affluent and strong. They can control the implementation of planning in such a manner that they receive lion share of planning. The scheduled castes, scheduled tribes and most of the muslims belong to the weaker section of the community and are almost deprived from getting such benefits.

Caste Structure and Utilisation of Water Resources

The objective of this part of study is to identity a relationship between distribution of castes and utilisation of water resources.

Inter Caste Relationship

Before going for the above it is necessary to examine the nature of co-relationship between population belonging to scheduled castes, scheduled tribes and general population (Table 14.1).

TABLE 14.1: Relationship between scheduled castes, scheduled tribes and general population, Distribution of 'V' value

	Scheduled Caste	Scheduled Tribes	General population
Scheduled Castes	1.00	.23*	.89**
Scheduled Tribes	.23*	1.00	.17
General Population	.89**	.17	1.00

N = 19, ** Significant at 0.01 level of significance.
* Significance at 0.50 level of significance.

The above Table 14.1 shows that in Birbhum district the scheduled castes population resides mainly with the general population, i.e. high caste Hindu, middle castes Hindu and Muslim population. Several social scientists have opined that the scheduled castes are functionally tied with the high and middle caste Hindu farmers. These land owners prefer to get their labourers and share croppers mainly from scheduled castes communities. That also indicates that both the general and scheduled castes population reside in those areas where agricultural development has taken place to some extent, whereas the backward places of western part of the district is the dwelling place of mainly scheduled tribes population.

Relation Between Population Belonging to Different Castes Groups and Irrigation

The study further indicates that the high, middle caste hindu and muslim population reside in large number in those blocks where both net and gross irrigated area are very high. The affluent people take the advantage of shallow tubewell, tank, river lift irrigation, and canals to irrigate their land.

The concentration of scheduled castes population is confined to the areas where large number of general population reside. So the number of Scheduled Castes population is much higher in the areas where net irrigated area, gross irrigated area, area irrigated by canal, river lift irrigation, tank are quite extensive. But the picture is just the opposite in case of scheduled tribes population. Although the irrigated area as a whole and under different sources are positively correlated in most of the cases, but that is the chance factor only. In case of larger observations the same pattern may not be evolved, because the correlation coefficient values are insignificant at any level of significance. So, the conclusion may be drawn, that the implementations of irrigation planning have so far been confined to those favourable areas where high and middle caste Hindu and Muslim population are concentrated. This area is a continuous patch consisting of northern, central and south eastern parts of the district.

TABLE 14.2: Relationship between castes structure and distribution of sourcewise irrigated areas.

(distribution of 'r' values)

Population	Net irrigated area	Gross irrigated area	Area Under				
			Canal	DTW	Shallow	RLI	Tank
General Population	.67**	.76**	.50	.29	.69**	.58	.60*
Scheduled Tribes Population	.34	.41	.13	.22	.07	.37	.29
Scheduled Castes	.81**	.79**	.68**	.26	.58	.64*	.56

N = 19, ** Significant at 0.01 level of significance
* Significant at 0.05 level of significance

Note: Less than .29 (r value) is insignificant.

Relation Between Farmers Belonging to Different Castes Groups and Net Irrigated Area and Net Value of Crops

It is worthwhile to study the distribution of net irrigated area and net value of crops in Rs. per acre of GCA as against the distribution of high and middle castes Hindu population, and distribution of S.C., S.T. populations at the village level. Fifteen villages have been selected randomly from southern part of the district.

The study reveals six kinds of relationships. These are give out of fifteen villages where large number of high and middle caste Hindu population reside. They take full advantage of different sources of irrigation in agriculture and raise higher productivity of crops with diversified and remunerative cropping pattern, hence can raise higher amount of net value of crops in rupees per acre of gross cropped area.

There are two villages out of fifteen, where large number of high and middle castes Hindu population reside. Inspite of low proportion of irrigated areas, those villages receive higher amount of net value of crops in rupees per acre of G.C.A. (Table 14.3). These villages are located at the low land part of the district where soil has higher water holding capacity, so that the soil is both fertile and productive. The productivity values of crops are quite high in these villages. The deficiency in irrigation is considerably met by the location of the villages at low land areas.

As against this, there is only one village located at the elevated part of the district. Here soil is porous, sandy-lateritic, and overdrained. Inspite of higher proportion of irrigated area, the village generated low amount of net value of crops. This village is inhabited by a large number of high and middle caste Hindu population. The terrain condition became the main hurdle against the proper use of technology.

TABLE 14.3: Percentage of net irrigated area, net value of crops in Rs. per acre of GCA and castes structure

Sl. No.	Name of Villages (Police Station)	% of NIA	Net value of crops/acre of GCA	% of high middle castes Hindu population	% of S.C. and S.T. population
1.	Jalandi (Nanoor)	70.02	1013.50	71.98	28.02
2.	Gomai (Nanoor)	99.16	911.78	83.05	16.92
3.	Kapastikuri (Bolpur)	87.68	1275.11	44.44	55.55
4.	Raipur (Bolpur)	86.53	1135.09	47.67	52.33
5.	Maheswarpur (Illambazar)	97.05	889.78	44.44	36.22
6.	Brahmanpara (Labpur)	46.81	1142.86	66.19	33.80
7.	Purba Sibpur (Khoyrasol)	25.70	1018.91	51.75	50.31
8.	Pansuri (Khoyrasol)	59.56	755.21	60.34	39.65
9.	Uchchkaran (Nanoor)	97.42	850.75	15.88	60.60
10.	Baitara (Nanoor)	68.93	878.19	0.63	14.79
11.	Baidyanathpur (Bolpur)	69.82	896.16	33.33	65.47
12.	Ushahar (Illambazar)	22.63	1154.30	8.42	92.08
13.	Khagra (Dubrajpur)	30.56	1077.56	28.35	71.65
14.	Jahanabad (Bolpur)	33.51	696.39	1.91	60.00
15.	Beledanga (Khoyrasol)	18.18	784.49	12.30	60.00

Source: Field investigation.

There are three villages which are located at the lowest part of the district in the south eastern corner — a flood-prone area. Besides surplus irrigation-water stagnate on these low lands for some days together, causing water loosing. These physical hindrances decline the productivity values of crops. These villages are inhabited by a large number of S.C. and S.T. population. Inspite of the available irrigation facilities, the poor farmers cannot make full use of it because of flood and waterlogging problems. These poor farmers can neither protect their fields from flood havoc, nor can reclaim them. That is why, these villages generate low amount of net value of crops per acre of GCA. So, the physical constraints intermingle with the poverty, illiteracy etc. throw adverse impact on productivity of crops.

Two villages are inhabited by mainly scheduled castes population, where cultivated areas are irrigated by deep and shallow tubewells. These tubewells although irrigate small portion of the gross sown area of the village, but due to intensification of crop on that small portion of land with the help of remunerative and commercialised cropping pattern, the villagers earned high amount of net value of crops. The villagers make serious effort and work hard with their own labour force and extract the land resource significants.

The last pattern includes two villages. These two villages are located at the lap of the river Ajoy, which are flooded almost every year. Poor farmers mainly reside there where unirrigated areas are available in large amount. Hence, generate low amount of net value of crops in rupees per acre of GCA. So, poverty, illiteracy, lack of technology, lack of human efficiency and physical constraints together give rise to low productivity of crops in these villages.

Above discussions leads to following conclusions.

(a) The villages, which have rolling plane terrain with minimum physical constraint and have favourable climate and, soil resources for agricultural development, have concentration of high and middle caste Hindu population who use appropriate irrigation technologies in the villages.

(b) The village with several physical constraints, neither have adequate irrigation facilities nor have extensive irrigated areas. The high middle and castes Hindu population mainly reside there.

(c) The villages with considerable environmental constraints have larger proportions of scheduled castes and scheduled tribes, population. These village have lower share of irrigated areas.

(d) The villages with negligible physical constraints have the large proportion of irrigation area. These villages are inhabited by poor farmers.

Relation Between Farmers Belonging to Different Castes Groups and Sourcewise Irrigated Areas

In this part an attempt is made to examine the extent to which the land of different caste groups of the study area receive irrigation from different sources.

From fifteen sample villages, three hundred and sixty households are selected randomly from different caste groups as indicated in the above table. These sample households of different caste groups may be expected to represent all the households of respective caste groups of the area. The relationship between different variables and the caste groups has been shown in Table 14.4.

TABLE 14.4: Caste groups and sourcewise distribution of irrigated area

Caste Groups	% of canal irrigated area	% of Tank irrigated area	% of DTW irrigated area	% of sha-llow area	% of R.L.I. irrigated area
High Caste Hindu	61.30	17.10	0.00	7.20	14.40
Middle Caste Hindu	45.23	25.48	16.37	2.17	10.45
Scheduled Castes	34.40	18.10	14.25	6.12	27.13
Scheduled Tribes	30.22	15.23	0.00	0.00	54.55
Muslims	36.12	21.18	0.00	26.36	16.34

Note: Percentages are calculated from the total of individual category and not from the total of the region.

Source: Field investigation.

The study envisages that the farmers belonging to the high and middle caste Hindu groups have larger share of irrigated land under canal, tank, shallow, and river lift irrigation schemes. The poor farmers' shares (scheduled castes, and scheduled tribes) are quite low, although it is not insignificant. The scheduled castes farmers irrigate large proportion of their landholding by deep tubewell and riverlift irrigation schemes. Muslim farmers utilise the groundwater resources through shallow tubewells, and have extensive irrigated areas under that scheme.

Canal, with flush irrigation can be irrigated vast stretches of land. Although canal cannot always satisfy the farmers by supplying irrigation during the time of need, yet the high and middle castes Hindu farmers can make maximum use of canal water, then the farmers of other castes.

Although most of the tanks of the region are derilict, yet these are used mainly to irrigate the winter and summer crops. The tank irrigation rescues the farmers when due to inadequate supply of canal irrigation, the standing crops are in serious condition. The high and middle caste Hindu farmers have larger shares of tank irrigated area.

Shallow tube-wells are either owned by high and middle caste Hindu farmers or by the Government. The government owned shallow tubewells are installed mainly to irrigated poor farmers' land, but these are not properly maintained. In many occasions the anti social activities, encouraged by the owners of shallow tubewells (land lords), destroy them, so that the poor farmers have no other alternative, but to purchase water from those land lords. On the other hand the tubewells, owned by big farmers irrigate land properly and are maintained suitably.

The pump sets or other water lifting devices are owned either by high and middle caste Hindu farmers or by the government. The government owned pump sets irrigate the land of all farmers, specially the weaker sections, but in many cases these are not maintained properly. On the other hand, the pumpsets owned by mainly the big farmers are maintained, get proper services of repair etc. These irrigate the land of the owners only.

It is believed that the district does not have adequate physical and socio-economic potentialities to irrigate substantial amount of land by deep tubewell. Only the middle caste Hindu and scheduled castes farmers get a large portion of their land irrigated by deep tubewells. Government owned deep tubewells are locally controlled, yet their operations are not free from biasness. Generally, they are operated to satisfy the interest of the "strong" farmers.

Conclusion

The above discussion leads to following conclusions:

(i) The high and middle caste Hindu population concentrate on the gently rolling plain topography of the district, where agricultural development has taken place to a certain level and has considerable scope for further development. Scheduled castes population are functionally tied with the above castes. They also reside in large number in the same areas. However, the scheduled tribes population are confined mainly to the undulating topography where meagre natural resources are available for agricultural development.

(ii) Irrigation planning follows general resource pattern of the region but it is quite different in prone areas of natural calamity. That means net irrigated areas are extensively found on the gently rolling plain surfaces where favourable natural resources are available, but the area is not frequently disturbed by natural calamities. Well to do farmers or poor farmers have made their residences in large number there, although it is found that the high and middle caste Hindu population mainly cluster in those areas.

(iii) The high and middle castes Hindu farmers have well control over all sources of irrigation, but the scheduled castes farmers have larger share of irrigated land that are irrigated by canal and river lift irrigation schemes. Most of the land of scheduled tribes are irrigated mainly by river lift irrigation devices and tanks. Muslims have higher proportion of land irrigated by canal and shallow tubewells. Deep tubewell is rarely found in the area.

REFERENCES

Beteille, A., 1983 : *The Idea of Natural Inequality and Other Essays*, Oxford University Press, Delhi.

Ghosh, S. N., 1981 : *District Census Hand Book, Birbhum District*, Part XIII-B, Director of Census Operations, West Bengal.

Minhas, B. S., 1978 : *Planning and the Poor*, S. Chand and Company Ltd., New Delhi.

Misra, P. K., 1982 : *Economic Development and Social Stratification*, D.K. Publications, Delhi.

Thorner, D. and A., 1974 : *Land and Labour in India*, Asia Publishing House, New Delhi.

15

CHANGING PATTERN OF IRRIGATION IN ANDHRA PRADESH

M. V. REDDY AND N. B. K. REDDY

IRRIGATION IS identified as a decisive factor in Indian agriculture due to high variability and inadequacy of rainfall. Irrigation is imperative for successful agriculture particularly in arid, semi-arid and sub-humid areas which are prone to drought and famine conditions due to partial failure and delayed arrival or early withdrawal of the monsoons. The present study area, Andhra Pradesh, has substantial area under semi-arid climate. Here, agriculture is a gamble with monsoon.

The Study Area

Andhra Pradesh is the biggest state in south India. It ranks fifth both in size (277000 km^2) and population 53.55 millions as per 1981 census) among the states of India. On the basis of relief and structure, the state can be divided into three major physical regions viz., (i) the coastal plain, (ii) the Eastern ghat region, and (iii) the interior plateaus.

The coastal plain is a monotonously flat plain except when it is broken across by the rivers and streams. It is widest in the middle where Krishna and Godavari rivers formed the deltas. The deltas and the adjacent plains constitute the agricultural heartland of the state. The Eastern ghat region comprises a series of broken hills and ridges generally running in north-south direction. The interior plateau extends largely between the contours 150-600 mts. This is an old peneplain, deeply dissected and scattered with hillocks of small relief. The slightly undulating topography has facilitated damming of rivers which are beaded with tanks along their courses (Table 15.1). Though Andhra Pradesh is called a state of rivers, only the Godavari and the Krishna are large perennial rivers, the Pennar is a large seasonal river, while the others are medium and minor rivers numbering over 35.

TABLE 15.1: Extent of river basins

	Name of the Basin	Area (in percentage of the total area of the state)
I.	*Major Basins* :	72.7
	1. Krishna Basin	27.4
	2. Godavari Basin	27.0
	3. Penna Basin	18.3
II.	*Medium and Minor Basins* :	27.3
	1. Vamsadhara and other coastal river basins (north of Godavari)	8.8
	2. Gundlakamma and others (between Krishna and Penna)	8.8
	3. Kolleru lake	4.3
	4. Others (south of Penna)	5.5

The state has a typical tropical climate and is marked for its seasonality. The four distinct climatic seasons identified are — (i) Summer season (March-May) (ii) The S.W. Monsoon season (June-September) (iii) The retreating monsoon season (October-November) and (iv) The Winter season (December-February). The south-west Monsoon season is the main rainy season for the entire state except for the south-eastern part. The mean annual rainfall ranges from less than 60 cm over drought prone arid regions of Rayalaseema (Southern interior plateau) to over 140 cm in northern parts of the state. Most of the interior parts receive less than 80 cm of rainfall. Much of the coastal belt receives over 100 cm. Almost the entire Telangana region (Northern interior plateau) gets over 75 per cent of its rainfall during this season. In many areas the proportion exceeds 90 per cent. The Rayalaseema and coastal areas receive 50 to 75 per cent of the mean annual rainfall during this season. The Rainfall from this monsoon decreases to the south from the Krishna river. South-eastern part receives only 25 to 50 per cent of its rainfall in this season. It receives more of its rainfall during the retreating and N.E. monsoon. For the year as a whole there is a net water deficit all over the state. The water deficit ranges from 600 mm in the coastal region to a maximum of 1100 mm in the driest parts of Rayalaseema. The least deficit is found in Krishna-Godavari deltas, in northern Telangana and in the Eastern hill region. Marked high deficit of rainfall is noticed in large parts of Rayalaseema and south-west Telangana. Further, most of the state is covered with red loamy soils which are of very low moisture retentive. The state is predominantly rural in character with 76.5 per cent of the population living in rural areas. Agriculture is the main occupation of the people with nearly 70 per cent of the working population engaged in agricultural sector.

Irrigation Potential and Development in the State

The three major rivers and over 35 medium and minor rivers drain the state of Andhra Pradesh. According to the second Irrigation Commission report 1972, the state has an aggregate irrigation potential of about 10.32 million hectares. It includes about 6.48 million hectares under major and medium irrigation, about 2.02 million hectares under minor irrigation and about 1.82 million hectares by ground water resources. An against this potential the gross irrigated area in 1972-73 was only 3.56 million hectares accounting for only 34.5 per cent of the potential. It increased to 4.56 million hectares in 1982-83 accounting for 44.2 per cent. More than 50 per cent of the irrigation potential is yet to be harnessed in the state.

Canal Irrigation

Andhra Pradesh ranks second among the states of the country in canal irrigation. It is a major type of irrigation in the state accounting for nearly 50 per cent of the gross irrigated area. Among the Irrigation projects in the state, Nagarjunasagar project on the river Krishna is the biggest one. The other important projects are Pochampadu or Sriramsagar on river Godavari, Tungabhadra Low level and High level canals, Kurnool-Cuddapah canal, Godavari barriage, Kaddam project and Vamsadhara project. As many as 54 major and medium irrigation projects have been developed in the state so far (Table 15.2).

Tank Irrigation

Tank irrigation is the second important type of irrigation in the state during 1972-73 but pushed to the third place during 1982-83. It accounts for over 25 per cent of the gross irrigated area in the state.

Agriculture under tank irrigation is relatively more risk-prone than under canal or well irrigation. Majority of the tanks are seasonal and rainfed in character. In Telangana the number of tanks are more or less evenly distributed. In Rayalaseema, the Chittoor district has the largest number of tanks. Tank irrigation is absent in the deltaic regions where assured canal irrigation is available (Table 15.3).

TABLE 15.2: Some major irrigation projects in Andhra Pradesh 1982-83

Name of the Project	Irrigated area (in '000 Hectares)	Districts Covered
Nagarjuna Sagar	676.5	Guntur, Krishna, Nalgonda, Prakasam and Khammam
Sreeram Sagar	151.0	Nizamabad, Karimnagar and Warangal
Kaddam	26.8	Adilabad
T.B.P.H.L.C. Stage I	47.9	Anantapur and Cuddapah
T.B.P.H.L.C. Stage II	29.0	Cuddapah
T.B.P.L.L.C.	60.3	Kurnool
K.C. Canal	112.5	Kurnool and Cuddapah
Vamsadhara	18.0	Srikakulam

TABLE 15.3: Region-wise distribution of irrigation tanks 1982-83

Region	Number of Tanks		Total
	Large Tanks	Small Tanks	
Coastal Andhra	2800	26553	29353
Rayalaseema	970	9562	10532
Telangana	3654	30534	34188
Total	7424	66649	74073

Note: Large Tanks : > 40 hectares of irrigation capacity
Small Tanks : ≤ 40 hectares of irrigation capacity

Well Irrigation

Utilisation of ground water at shallow depths by means of irrigation wells is an important method of irrigation in the non-deltaic tracts. Introduction of oil engines and electric pumpsets in place of traditional methods of water lifting, has helped rapid development of well irrigation in the last two decades. Though well irrigation is wide spread in the state, it is a major source of irrigation particularly in Karimnagar and Warangal districts of Telangana and Chittoor District of Rayalaseema region. Tubewell irrigation is mostly confined to the coastal districts. During the year 1982-83 well irrigation accounted for 26.8 per cent of the gross irrigated area in the state relegating tank irrigation to third place (Table 15.4).

TABLE 15.4: Distribution of irrigation wells in Andhra Pradesh 1982-83

Name of the Region	Number of Tube Wells and Filter Points	Number of other Wells	Total Number
Coastal Andhra	66867	132830	199697
Rayalaseema	1005	215695	216695
Telangana	2750	498805	501555
Total	70622	847330	917952

Regional Variations in Irrigation Development

Canals, tanks and wells together account for more than 95 per cent of the irrigated area. The remaining area is under other sources which mostly include natural springs. During 1982-83 canal irrigation was found in 137 total taluks accounting for 49.1 per cent of the total irrigated area. Well irrigation was found in 181 taluks accounting for 26.8 per cent of the irrigated area. Area under tank irrigation decreased substantially between 1972-73 and 1982-83. During the period tank irrigation decreased (Table 15.5 and 15.6).

TABLE 15.5: Distribution of irrigation type

Type of Irrigation	1972-73	1982-83
	No. of Taluks	No. of Taluks
Canals	132	137
Wells	185	181
Tanks	184	172
Others	142	141

Note: Total number of Taluks : 195

TABLE 15.6: Extent of irrigated area under each type

(in lakh hectares)

Type of Irrigation	1972-73		1982-83	
	Area	% to Total	Area	% to Total
Canals	17.44	49.0	22.35	49.1
Wells	7.58	21.4	12.20	26.8
Tanks	8.90	25.0	9.00	19.8
Others	1.66	4.6	1.97	4.3
Andhra Pradesh Total	35.58	100.0	45.52	100.0

From 25.0 per cent to 19.8 per cent of the gross irrigated area. However the absolute area under tank irrigation showed a slight increase from 8.90 lakh hectares to 9.0 lakh hectares during the period. It is clear from the data that over 75 per cent of the canal irrigated area is concentrated in coastal Andhra. Telangana region showed an increase in canal irrigated area from 12.8 per cent in 1972-73 to 15.4 per cent in 1982-83, while Rayalaseema recorded a decrease.

There is not much change in the well irrigated areas. But area under tank irrigation showed substantial spatial change. Both Coastal Andhra and Rayalaseema relatively suffered substantial decrease while Telangana registered relatively substantial increase between 1972-73 and 1982-83.

Region-wise analysis of the irrigated area under different types shows that between 1972-73 and 1982-83 Coastal Andhra recorded an increase in irrigated area under all categories (Table 15.7 and 15.8). Canal irrigation plays a crucial role in this region accounting for over 65 per cent of the irrigated area (Table 15.9). During this period Rayalaseema recorded a slight fall in irrigated area. In this region canal, well and tank irrigations were almost of equal importance during 1972-73. But area under tank irrigation decreased substantially during 1982-83 (Table 15.8). Telangana region witnessed substantial increase in all categories of irrigation during this period. The three principal types are more or less equally important, though well irrigation is slightly dominant.

TABLE 15.7: Irrigated area under each type

(in percentage)

Type of Irrigation	Coastal Andhra		Telangana		Rayalaseema	
	1972-73	1982-83	1972-73	1982-83	1972-73	1982-83
Canals	76.8	75.8	12.8	15.4	10.4	8.9
Wells	32.3	31.2	42.9	43.3	24.8	23.5
Tanks	55.4	48.2	22.1	41.8	22.5	10.0
Others	31.0	37.5	12.0	52.9	56.0	9.6
Total	59.8	57.3	21.5	29.7	18.6	13.0

TABLE 15.8: Region-wise irrigated area under each category

(in lakh hectares)

Type of Irrigation	1972-73		1982-83	
	Area	%	Area	%
Coastal Andhra	21.27	100.0	25.79	100.0
Canal	13.39	62.9	16.91	65.6
Wells	2.44	111.5	3.81	14.8
Tanks	4.92	23.2	4.33	16.8
Others	0.52	2.4	0.74	2.9
Rayalaseema	6.63	100.0	5.96	100.0
Canal	1.82	27.5	2.00	33.6
Wells	1.88	28.4	2.87	48.1
Tanks	2.00	30.1	0.90	15.1
Others	0.93	14.1	0.19	3.2
Telangana	7.65	100.0	13.53	100.0
Canal	2.23	29.1	3.44	25.4
Wells	3.25	42.4	5.29	39.0
Tanks	1.97	25.8	3.76	27.8
Others	0.20	2.7	1.04	7.8

Analysis of Irrigation Pattern and Types

In the present study the distributional aspects of irrigation in Andhra Pradesh is analysed by means of three different approaches.

(i) The spatial variation of irrigation intensity is studied using the following index.

$$\text{Irrigation Intensity} = \frac{\text{Gross Irrigated Area}}{\text{Gross Cropped Area}} \times 100$$

(ii) The concentration aspect of each type of irrigation is studied by applying Florence location quotient index formula (1959).

$$L_{wi} = \frac{W_{ij}}{\sum_{j=1}^{m} W_{ij}} \Bigg/ \frac{\sum_{i=1}^{n} W_{ij}}{\sum_{i=1}^{n}\sum_{j=1}^{m} W_{ij}}$$

The above formula is modified and rewritten as follows for convenience of computation.

$$L_{wi} = \frac{W_{ij}}{\sum_{i=1}^{n} W_{ij}} \Bigg/ \frac{\sum_{j=1}^{m} W_{ij}}{\sum_{i=1}^{n}\sum_{j=1}^{m} W_{ij}}$$

where

L_{wi} = Location quotient

W_{ij} = Number of hectares in '*i*'th category of irrigation in '*j*'th taluk

$\sum_{j=1}^{m} W_{ij}$ = Number of hectares in '*i*'th category of irrigation in all taluks.

$\sum_{i=1}^{n} W_{ij}$ = Number of hectares in all categories of irrigation in '*j*'th taluk

$\sum_{i=1}^{n}\sum_{j=1}^{m} W_{ij}$ = Number of hectares in all the categories of the irrigation in all taluks.

(iii) The orientation aspect of the irrigation is studied employing Kostrowicki's method.

Intensity of Irrigation

In the present study the intensity of irrigation is expressed as the percentage of gross irrigated area to the gross cropped area (Fig. 15.1). This measure in a way gives the degree of intensiveness as well as the dependability of agriculture.

TABLE 15.9: Region-wise irrigation intensity

(In lakh hectares)

Region	1972-73			1982-83		
	G.C.A	G.I.A	G.I.A (as percentage G.C.A.)	G.C.A	G.I.A	G.I.A (as percentage of G.C.A.)
Coastal Andhra	46.23	21.28	46.0	55.24	26.16	47.4
Rayalaseema	30.77	6.63	21.5	29.88	5.95	19.9
Telangana	51.94	7.65	14.7	63.61	13.54	21.3
Andhra Pradesh	128.94	35.56	27.6	148.73	45.62	30.7

Note: G.C.A. : Gross Cropped Area; G.I.A. : Gross Irrigated Area.

The average intensity of irrigation was only 27.6 per cent in the state during 1972-73. But regionally it ranged from as low as 14.7 per cent in Telangana to 46.0 per cent in Coastal Andhra. The gross irrigated area increased from 27.6 lakh hectares in 1972-73 to 45.65 lakh hectares in 1982-83. Correspondingly the irrigation intensity also increased to 30.7 per cent in the state. But during the same period Rayalaseema recorded a decrease in the intensity while both Coastal Andhra and Telangana recorded marginal increase (Table 15.10). The decrease in Rayalaseema region is mainly due to fall in area under tank irrigation.

TABLE 15.10: Distribution of irrigation intensity

Intensity	Coastal Andhra	Rayalaseema	Telangana	Total
	No. of Taluks	No. of Taluks	No. of Taluks	No. of Taluks
1972-73	79	42	74	195
<20	10	17	37	64
20-40	22	17	28	67
40-60	25	6	7	38
60-80	8	1	1	10
>80	14	1	1	16
1982-83	79	42	74	195
<20	20	21	49	90
20-40	27	15	18	60
40-60	24	5	4	33
60-80	5	—	3	8
>80	3	1	—	4

In coastal Andhra the intensity of irrigation is very high (>80%) in 14 taluks and high (60-80%) in 8 taluks during 1972-73. The intensity is very low (<20%) only in 10 taluks mostly situated in the northern hilly region. The rest of the coastal region mostly enjoyed low to moderate intensity of irrigation.

An analysis shows a marked decrease in the intensity in some areas during 1982-83. As many as 47 taluks had low to very low intensity as against 32 taluks while only 8 taluks had high to very high intensity as against 22 taluks during 1972-73 in coastal Andhra. High and very high irrigation intensity was mostly found in the deltaic taluks of the middle coastal Andhra and a few taluks in the southern Andhra. The intensity of irrigation is very low in Rayalaseema region during 1972-73, with 17 taluks having very low (<20%) intensity and another 17 taluks having low intensity (20 to 40%). Moderate intensity (40 to 60%) was found only in 6 taluks mainly lying along the K.C. canal. In 1982-83 this region also witnessed a decrease in the irrigation intensity with 36 taluks having low to very low intensity as against 34 taluks in 1972-73. But basically the spatial pattern of intensity remained the same. In Telangana region the intensity is very low in 37 taluks and low in 28 taluks in 1972-73. The corresponding figures for 1982-83 are 49 and 18 taluks respectively. Very high to high irrigation is found only in 2 taluks in 1972-73 and 3 taluks in 1982-83. Though the extent of irrigated area is increasing, the increase is not in pace with the increase in cropped area. Besides in some areas dry farming is adopted in one season with the result that the intensity of irrigation has shown a decrease.

Concentration of Irrigation Types

Though all the important types of irrigation are present in many of the taluks, the concentration of different categories substantially varied from taluk to taluk. In some of the taluks some types of irrigation have only a nominal presence, while in others they account for almost 100 per cent of the irrigated area. The study of concentration aspect brings out clearly the areas having more/less than their legitimate share of particular type of irrigation. If the total irrigated area under particular category is evenly distributed the index value is unity. If a region has a location quotient index more than one, it has a higher concentration of that type of irrigation and if the index value is less than one the concentration is low.

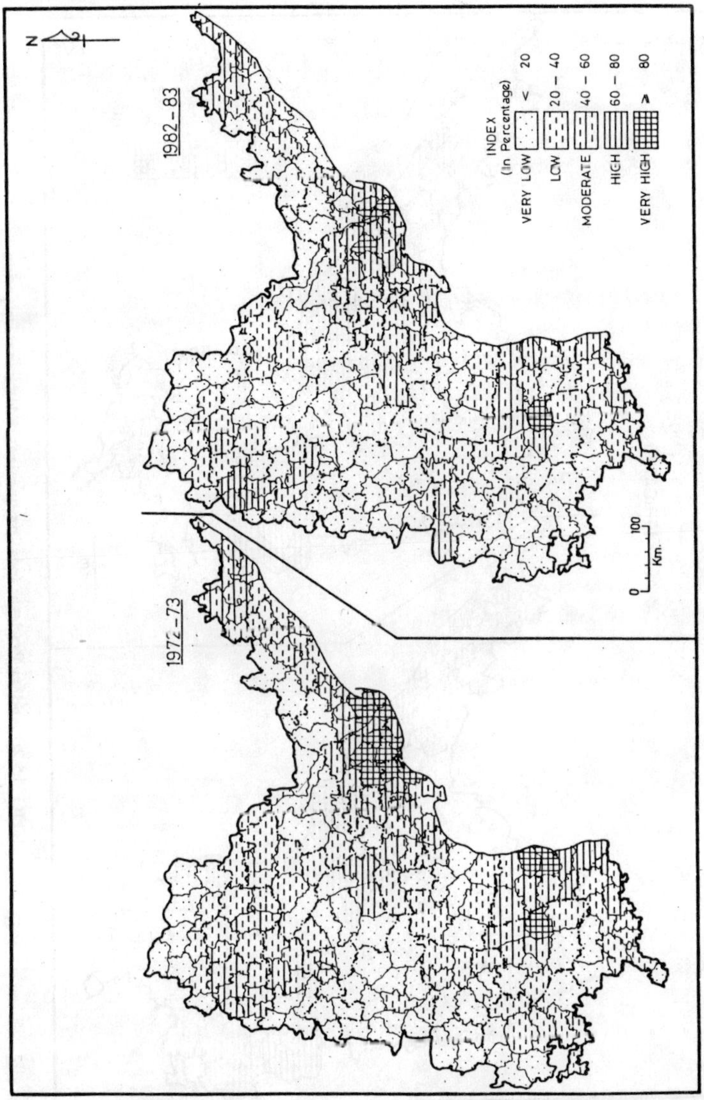

FIG. 15.1: Andhra Pradesh: intensity of irrigatio..

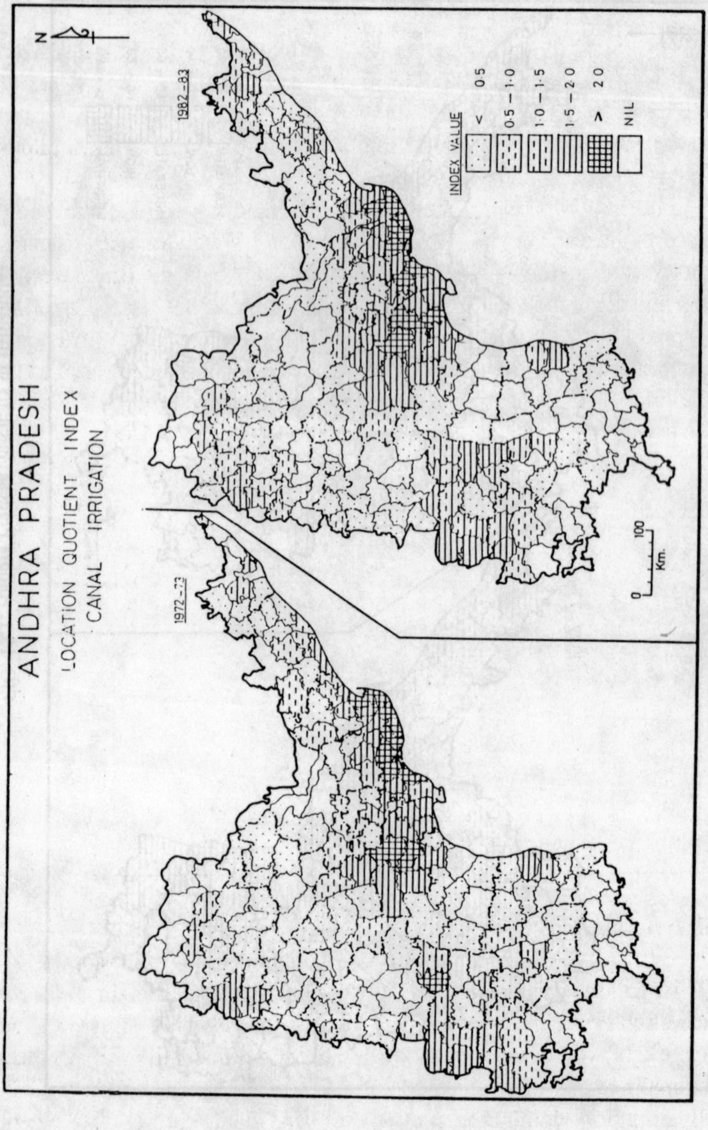

FIG. 15.2: Andhra Pradesh: location quotient index canal irrigation.

Canal Irrigation

Canal irrigation, though accounts for nearly 50 per cent of the irrigated area, it is found only in 70 per cent of the taluks in 1982-83. Most of this area is in Krishna and Godavari deltas. During 1972-73 canal irrigation was distributed in 132 taluks (Table 15.11). But in 55 taluks the concentration is very low (<0.5). Very high (>2.0) and high (1.5 to 2.0) concentrations were found in 36 taluks. Most of them were situated in Guntur, Krishna. East Godavari and West Godavari districts of the coastal region and a few along the K.C. canal of Rayalaseema region and Nizamabad of Telangana region. In 1982-83 canal irrigation has spread to 137 taluks. But the major areas of canal irrigation concentration remained the same. The major areas where canal irrigation is absent are the hilly Eastern ghat region, and a few pockets in the dry interior plateau region (Fig. 15.2).

TABLE 15.11: **Concentration of canal irrigation**

Location Quotient Lwi	1972-73	1982-83
	No. of Taluks	No. of Taluks
< 0.5	55	48
0.5 - 1.0	21	26
1.0 - 1.5	20	22
1.5 - 2.0	23	32
> 2.0	13	9
NIl	63	58
Total	195	195

Well Irrigation

Well irrigation was present in 95 per cent of the taluks in 1972-73 which slightly shrinked to 93.0 per cent of the taluks in 1982-83. The few pockets where well irrigation is absent are the hilly Paderu, and Chintapalli taluks of Vishakhapatnam district and few deltaic taluks. Well irrigation dominates the whole of Telangana region (except Adilabad district), western parts of Nellore and Prakasam districts of Coastal Andhra and the entire Rayalaseema except in a few pockets, where canal irrigation is dominant. There is not much change in the pattern of concentration between 1972-73 and 1982-83 except in a few taluks in Nalgonda, Krishna and Nizamabad districts, where canal irrigation is recently developed. In 1972-73 and 1982-83 as many as 68 and 65 taluks had very high (>2.0) concentration of well irrigation (Fig. 15.3). These taluks were situated mostly in the drought prone areas. This clearly indicated heavy dependence of agriculture on well irrigation in these regions (Table 15.12).

TABLE 15.12: **Concentration of well irrigation**

Location Quotient	1972-73	1982-83
	No. of Taluks	No. of Taluks
< 0.5	58	55
0.5 - 1.0	23	22
1.0 - 1.5	20	14
1.5 - 2.0	16	25
> 2.0	68	65
NIl	10	14
Total	195	195

Tank Irrigation

Tank irrigation was present in 94.4 per cent of the taluks in 1972-73, but slightly shrank in number to 88.2 per cent of the taluks in 1982-83. In both the years very high (>2.0) concentration of tank irrigation is noticed in 61 taluks. Northern and southern parts of Coastal Andhra, South-western parts of Rayalaseema and northern parts of Telangana region exhibit high concentration during both periods (Table 15.13). Tank irrigation is absent mainly in the deltaic regions of central Coastal Andhra (Fig. 15.4).

TABLE 15.13: **Concentration of tank irrigation**

Location Quotient (Lwi)	1972-73	1982-83
	No. of Taluks	No. of Taluks
< 0.5	61	61
0.5 - 1.0	28	30
1.0 - 1.5	23	24
1.5 - 2.0	21	19
> 2.0	43	38
NIl	11	23
Total	195	195

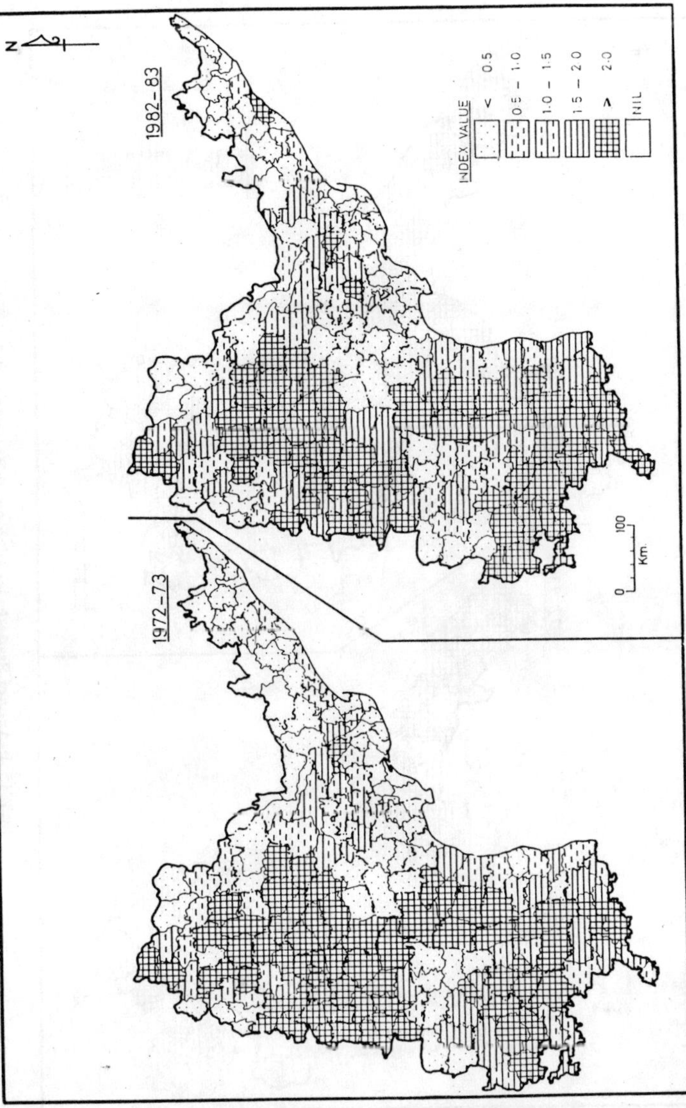

FIG. 15.3: Andhra Pradesh: location quotient index well irrigation.

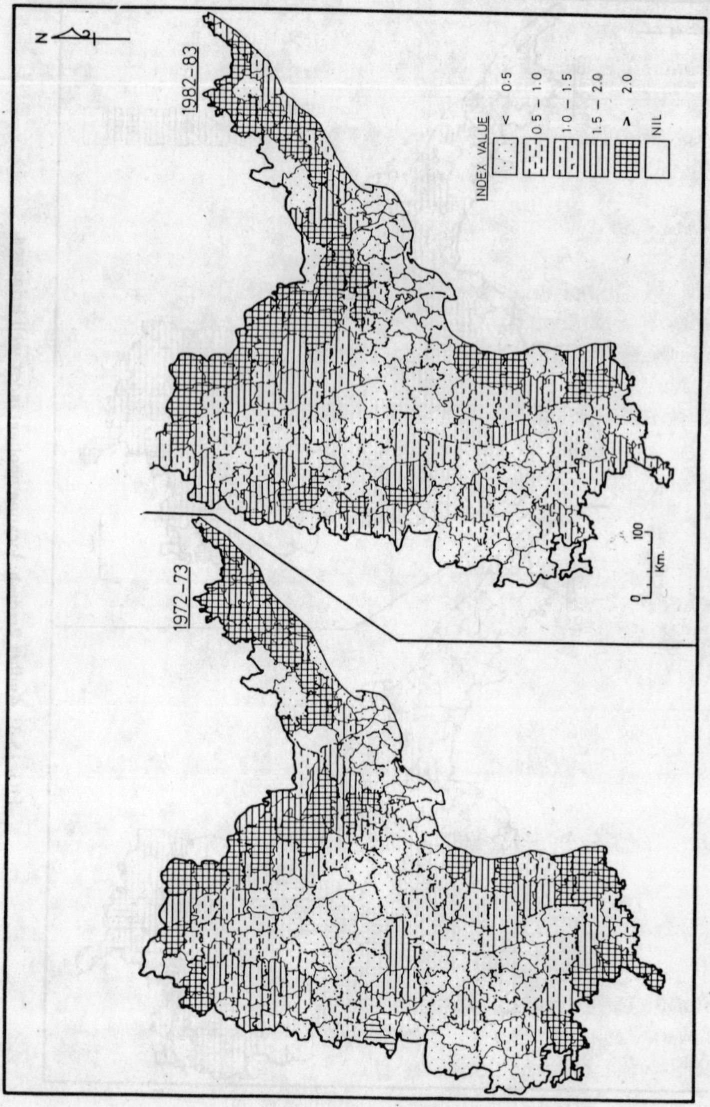

FIG. 15.4: Andhra Pradesh: location quotient index tank irrigation.

There is not much change in the spatial pattern of concentration during 1972-83 period except that in a few pockets in Telangana region like Warangal, Nizamabad and Khammam there is a marked change towards higher concentration in the tank irrigation, while in many parts of Rayalaseema region there is marked shift towards lower concentration.

Other Sources

Irrigated area under other sources accounts for less than 5 per cent of the gross irrigated area (Fig. 15.5). Its concentration is very high particularly in the hilly tracts where both perennial and non-perennial natural springs are present (Table 15.14). The important regions of high concentration are the hilly regions like Adilabad, Khammam, Vishakhapatnam and Chittoor Districts, and the deltaic parts like east and west Godavari, Krishna and parts of Nellore district and in some river valley pockets of Kurnool district.

TABLE 15.14: Concentration of irrigation under other sources

Location Quotient	1972-73	1982-83
	No. of Taluks	No. of Taluks
< 0.5	63	48
0.5 - 1.0	24	32
1.0 - 1.5	10	15
1.5 - 2.0	12	14
> 2.0	33	30
NIl	53	54
Total	195	195

Orientation of Irrigation

The orientation of irrigation is studied using Kostrowicki's method (1964, 1970) originally used for landuse orientation studies (Table 15.15).

The irrigation type having less than 20 per cent of the irrigated area is ignored. A symbolic nomenclature is given to the orientation of irrigation using the first letters of the irrigation category (Table 15.16). For example 'C' for canal irrigation 'W' for well irrigation and 'T' for

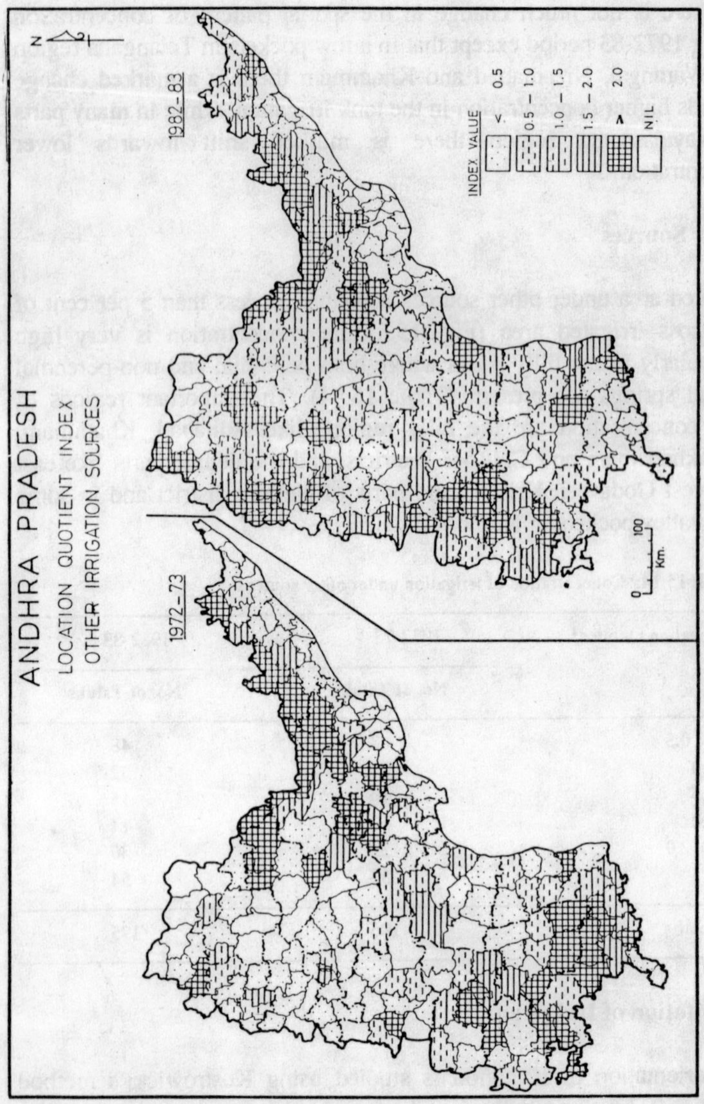

FIG. 15.5: Andhra Pradesh: location quotient index other irrigation sources.

TABLE 15.15: Kostrowicki's classification

Area (%)	Status of Irrigation	Score
> 80	Predominant	4
60 - 80	Dominant	3
40 - 60	Equidominant/Semi dominant	2
20 - 40	Accompanying	1

TABLE 15.16: Orientation of irrigation

Canal Based		Tank Based		Well Based	
Orientation	No. of Taluks	Orientation	No. of Taluks	Orientation	No. of Taluks
(1)	(2)	(3)	(4)	(5)	(6)
1972-73	63		62		65
C4	35	T4	18	W4	18
C3	5	T3	8	W3	5
C3 W1	3	T3 W1	13	W3 T1	16
C2	2	T3 C1	5	W3 C1	3
C2 W2	4	T2	2	W2	4
C2 T1	2	T2 W2	3	W2 T2	3
C2 W1	4	T2 C2	4	W2 C2	2
C2 T1 W1	2	T2 C1	3	W2 T1	9
C1 T1	2	T1 C1 W1	2	W2 C1	1
C1 T1 W1	1	T1	3	W2 C1 T1	2
1982-83	74		53		68
C4	36	T4	10	W4	24
C3	15	T3	8	W3	5
C3 T1	3	T3 W1	9	W3 T1	19
C3 W1	5	T3 C1	1	W3 C1	2
C2 T2	2	T2	2	W2	1
C2 W2	1	T2 W2	5	W2 T2	5
C2 T1	3	T2 W1	10	W2 C1	2
C2 W1	1	T2 C1	2	W2 T1	4
C2 T1 W1	1	T1 W1	2	W2 T1 C1	2
C1	2	T1	2	W2 C2	2
C1 T1	1			W1 C1 T1	1
C1 W1	1			W1 T1	1
C1 T1 W1	3				

tank irrigation. They are grouped into a formula where letters represent the irrigation category and figures represent their score (Fig. 15.6). The study is carried out for both 1972-73 and 1982-83. The study shows that the state as a whole had a semi dominant canal based irrigation with tank and well irrigation in accompanying status (C2+T1+W1) in 1972-73. But it changed to semi-dominant canal based irrigation with only well irrigation in the accompanying status (C2+W1) in 1982-83. It gives an indication of a change towards more dependable canal and well irrigation.

Canal Based Orientation

Canal based orientation is found in 63 taluks in 1972-73. In 35 of them it occupied a predominant place. It occupied dominant place in 8 taluks and semidominant or equidominant with/without canal or well irrigation accompanying in 16 taluks. It figured in accompanying status only in 3 taluks. During 1982-83 canal based orientation is found in 74 taluks of which it occupied a predominant status (C4) in 36 taluks and dominant status (C3) in 15 taluks having majority of them in the deltaic region. It figured in equidominant or semidominant position accompanying well or tank irrigation in 16 taluks, situated mostly in the central part of Rayalaseema and northern part of Telangana. It occupied accompanying status in only 7 taluks most of them situated in western Rayalaseema and Telangana regions where canal irrigation has been developed recently. The pattern of orientation clearly shows that in most of the canal based area, canal irrigation mostly assumes either predominant or dominant position.

Well Based Orientation

Well irrigation based orientation is found in 65 taluks in 1972-73 which increased to 68 taluks in 1982-83. In both the years almost similar spatial pattern is noticed. Well irrigation is predominant in 18 taluks and dominant in 5 taluks. Pre-dominant and dominant well irrigation is found mostly in the central parts of the Telangana region. Dominant well irrigation accompanied by tanks (W3 T1) is the other important orientation mostly found in southern part of Rayalaseema, southern and eastern Telangana regions. In most of arid regions semidominant and equidominant well irrigation is found accompanied by tank irrigation (W2 T2 or W2 T1) and only in a few regions by canal irrigation (W2 C1).

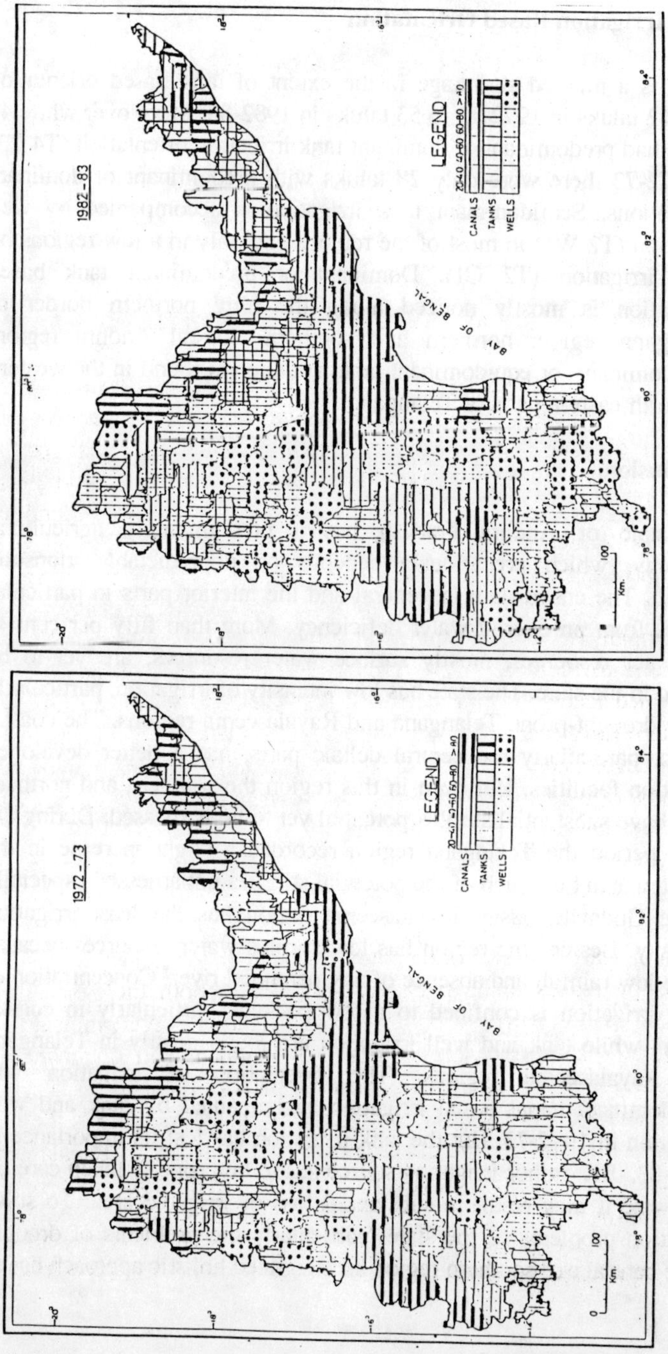

FIG. 15.6: Andhra Pradesh: orientation of irrigation.

Tank Irrigation Based Orientation

There is a marked shrinkage in the extent of tank based orientation from 62 taluks in 1972-73 to 53 taluks in 1982-83. More over while 44 taluks had predominant or dominant tank irrigation orientation (T4 T3) in 1972-73 there were only 28 taluks with predominant or dominant orientations. Semidominant tank irrigation is accompanied by well irrigation (T2 W1) in most of the regions and only in a few regions by canal irrigation (T2 C1). Dominant or predominant tank based orientation is mostly noticed along the hilly northern border of Telangana region, northern and southern Coastal Andhra region. Semidominant or equidominant tank irrigation is found in the western and north-eastern parts of Telangana.

Conclusion

The state of Andhra Pradesh has a predominantly agricultural economy, which precariously depends on unpredictable monsoon rainfall. The entire state in general and the interior parts in particular suffers from an overall water deficiency. More than fifty per cent of the water resources, mostly surface water resources, are yet to be tapped in the state. The state has low intensity of irrigation, particularly in the drought-prone Telangana and Rayalaseema regions. The coastal region, particularly the central deltaic parts, has a better developed irrigation facilities. But even in this region the southern and northern parts have substantial surplus potential yet to be harnessed. During the study period the Telangana region recorded a slight increase in the irrigated area but much of the potential still lies unharnessed especially in the Godavari basin. Rayalaseema region has the least irrigation intensity. Besides the region has low surplus water resources because of the low rainfall and absence of any perennial river. Concentration of canal irrigation is confined to a few pockets, particularly in coastal region, while tank and well irrigation are found mostly in Telangana and Rayalaseema regions. The orientation of irrigation was semidominant canal based irrigation accompanied by tank and well irrigation in 1972-73. But the tank irrigation has lost its importance in 1982-83. This minor irrigation source has to be properly taken care of, otherwise it may result in substantial loss of irrigated area. To solve irrigation problems of the state, particularly the problems of drought prone central and southern region an integrated holistic approach has to be adopted. As the surface water resources are unequally distributed in favour of the northern region, basin-wise planning for tapping the water resources may not solve the problems of the central and southern region. Maximum justice both to the region which has large water resources and to deficit parts of the state could be done if the three river basins are integrated into one system through inter-basin transfer of water. Besides, a judicious strategy has to be developed for conjunctive use of surface and ground water resources.

REFERENCES

Chaturvedi, B. N. and Reddy K. S., 1964 : "Irrigation in Telangana, Sources and Problems", *The Deccan Geographer*, Vol. II, No. 2, pp. 154-70.

Chaturvedi, B. N., 1968 : "The Origin and Development of Tank Irrigation in Peninsular India", *The Deccan Geographer*, Vol. VI, No. 2, pp. 57-86.

Florence, 1959 : Investment, Location and size of Plant, Cambridge University Press, Cambridge.

Kostrowicki, J., 1964 : Land Utilisation Survey as a basis for Geographical Typology of Agriculture, *Przeglad Geograficzing*, 32, pp. 169-83.

Ramanaiah, Y. V., 1985 : "Irrigation in Andhra Pradesh" Andhra Pradesh Economic Association Third Annual Conference Publication, 19-20 January.

———, 1974 : Andhra Pradesh Planning Atlas, Department of Geography, Osmania University, Hyderabad.

———, 1972 : Report of the Second Irrigation Commission, Vol. 1, 2 and 3, Government of India.

———, Seventh Five Year Plan, 1985-90, Andhra Pradesh, Vol. I & II, Finance and Planning Department, Government of Andhra Pradesh, Hyderabad.

———, 1961 : Techno Economic Survey of Andhra Pradesh, NCAER, New Delhi.

16

PROGRESS OF IRRIGATION IN UTTAR PRADESH PLAINS*

O. COUTINHO AND T. C. SHARMA

THE STATE of Uttar Pradesh has made great strides in irrigation in recent years. It has now over 53 per cent of the net sown area under irrigation, next only to Punjab, Haryana and Tamil Nadu. However, still much needs to be done in this regard. Large tracts still remain unirrigated. Bundelkhand and Central region bear testimony to the state of neglect. Besides haphazard and unplanned growth has compounded the problem all the more. This is borne out by widespread water-logging and salinisation of soils in canal irrigated areas and drying up of tubewells due to a falling water-table in tubewells irrigated areas all over the State. It is, therefore, high time to formulate well-conceived long term perspective irrigation development plans so as to utilise vast surface and groundwater resources of the States. Such measures can help in reducing the inter-regional growth disparity and adopting a technology free from the above referred to ill effects. Before that, it is essential to analyse in depth these problems in a spatio-temporal context which is by and large lacking so far.

The present study aims at (i) analysing the disparities in the growth of irrigation source-wise and in net and gross irrigated area between and within the five agro-climatic regions of Uttar Pradesh, (ii) focussing upon the problems resulting due to the haphazard and unplanned manner of irrigation facilities development, and (iii) suggesting measures necessary for a fuller exploitation of the State's potential surface and groundwater resources necessary for a balanced

* *Source: Agricultural Situation in India*, Directorate of Economics and Statistics Department of Agriculture and Co-operation Ministry of Agriculture, Feb., 1987.

and faster growth of agricultural production in all parts.

The study covers the period from 1966-67 to 1980-81. The selection of the year 1966-67 is based on the presumption that a rapid increase in irrigation began with the launching of the HYV Programme. The study area includes all the 48 districts of Uttar Pradesh existing in 1966-67. It excludes the North-western hill districts, except Dehradun and Nainital, for want of relevant data. Data for new districts formed after 1966-67 have been merged with the parent districts.

Unexploited Potential

According to the 1972 Irrigation Commission Report, Uttar Pradesh has the largest surface and groundwater resources among all the Indian States. These were estimated by the Commission at 151 lakh hectare-metres including 95.45 lakh hectare-meters in the form of surface water and 55.55 lakh hectare-meters as groundwater. These resources are considered adequate to provide protective irrigation to over 87 per cent of the State's net sown area of about 168 lakh hectares. The gross irrigated area in the State currently stands at about 109 lakh hectares of 47 per cent of the total cropped area. It is nearly 24 per cent of India's gross irrigated area and comparatively much larger than that of any other State. Uttar Pradesh, however, lags behind Punjab and Haryana in gross irrigated per cent, proportion of total cropped area, being 85.47 per cent and 60.58 per cent, respectively. A low incidence of multiple cropping and an error in the data (the State's large area under sugarcane, a heavily irrigated 10-12 months' crop, being counted only once) are responsible for the lower proportion in Uttar Pradesh.

Source-wise Development Trends

The irrigation sources in the study area fall into four traditional types: tubewells and other wells, canals, tanks and other sources. The first two account for nearly 95 per cent of the net irrigated area. Tubewells alone account for 52 per cent and have been mostly responsible for the rapid spread of irrigation facilities in recent years. Canals come next with about 33 per cent; they are mainly government-built and controlled. Other wells account for about 10 per cent. The other sources including tanks are important in only small localities and together irrigate only about 5 per cent of the irrigated area in the State.

Rapid Expansion of Tubewell Irrigation

Tubewell irrigation represents a modern, scientific and efficient method of exploiting groundwater and has greatly contributed to the success of the new agricultural technology, particularly in the Northern alluvial plains of India, including the study area, where tubewells now form the most important irrigation source with 46.95 lakh hectares or about 52 per cent of the State's net irrigated area. Well irrigation has been practised in Uttar Pradesh since ancient times and traditional lifts like *dhenkali, charsa* and *reht,* based on human and animal engery, were used for the purpose till recently. In case of a tubewell, the energy used is electrical power or diesel-motors, which lift groundwater through suction from deep bore-holes. Compared to any other lift, a tubewell yields more water and at a faster rate (25,000-30,000 litres per hour as against only 10,000 litres per hour from a *reht* well). A deep tubewell, tapping an acquifer at 100-500 metres depth, may yield on an average 1,35,000 litres of water per hour, enough to perennially irrigate 80-100 hectares of crop land. It may run for 2000-3000 hours per annum and have a life of 20-25 years. The minimum distance between the two nearest neighbour tubewells should be around 2 km. to maintain a sustained supply in case of a 62 metres-deep tubewell. These standards are, however, often flouted, which leads to a fall in the level of the water-table and drying up of neighbouring tubewells with a shallow depth.

Tubewells technology is found ideal in Uttar Pradesh plains due to a huge amount of groundwater found here in a series of sandy acquifers, separated by clay lenses, which yield a sustained supply of sweat water, especially when a dug well is strengthened with masonry work. Through the bottom of its pit pipes are lowered to interconnect two or more such water-bearing starta. Underground water reservoirs are replenished annually through infiltration of rain (20-25%), seepage from canals and rivers, and percolation of irrigation water from fields. It is estimated that of the 55.5 lakh hectare metres of groundwater resource found in the region, only 27 per cent was being exploited till 1970-71, the percentage of exploitation of the Western region being 36; Eastern Region 24; and Central region only 11.

Uttar Pradesh was a pioneer State in the use of tubewell technology in India. Construction of tubewells started here in 1931, when 1,500 tube-wells were installed in Meerut and Rohilkhand divisions, many were initially dug to augment water in the Upper Ganga Canal (Rao, 1975). Another important stage in its development was reached in 1952, when an agreement between the Government of India and the US Aid authorities was signed for the construction of 2,000 tubewells in Punjab, Uttar Pradesh and Bihar. The importance of tubewell irrigation was fully realised by the late Fifties by which time, private tubewells became a common slight in the North-western districts. The construction of a private tubewell using electricity/diesel is found relatively cheaper than a public tubewell due to its shallow depth and flexibility in terms of capacity and equipment. It works more efficiently where timeliness is concerned, and irrigation from shallow private tubewells is also cheaper in comparison to State tubewells, (Mellor and Moorti, 1971, Moorti, 1971, Moorti and Mellor, 1973, Mukhopadhayay, 1973 and Patel and Patel, 1969).

Real breakthrough in private tubewell irrigation was made after 1966-67 with the advent of high-yielding varieties. Possibilities of multiple cropping and incentive-oriented price policy too helped a great deal in the development of private tubewells. The farmers thought it prudent to have a source of irrigation under their command so as to be able to provide irrigation at the right time in right quantities. Between 1966 and 1980, the number of tubewells increased from a mere 5,000 to nearly 13 lakh or 260 times, private tubewells accounting for most of this increase. The tubewell irrigated area went up from 13.11 lakh hectares to 46.95 lakh hectares, recording a 358 per cent increase. It was merely 21 per cent of the net irrigated area in 1966-67 but rose to nearly 52 per cent by 1980-81. The new technology, as is well-known, opened new vistas of prosperity for the farmer and timely irrigation was a *sine qua non* for it, which should explain the collinearity between the area irrigated by private tubewells and the area under HYVs. The two specific programmes which provided a tremendous boost to the growth of tubewell irrigation in the region during the late Sixties were: (i) rural electrification, and (ii) provision of liberal capital finance through institutional sources on a large scale (Dhawan, 1975)

Thakur and Kumar (1984) using farm-level data of Meerut district for the year 1980-81, showed that yield rates, area under sugarcane and wheat (surrogate for commercialisation) and cropping intensity were much higher on private tubewell irrigated farms compared to government tubewell and canal irrigated farms. It is well known that farms using tubewell and canal irrigation in conjunction give the highest returns provided several other important conditions are also met with.

Tubewell irrigation, on account of a relatively large investment required for it, has brought into play the role of capital finance for the first time in Indian agriculture and it has become a source of growth in income disparity due to its multiplier effect in the economy. The contrasts in tubewell irrigation in the three plain regions are also due to several physical and cultural differences, e.g., rainfall distribution, floods, depth of groundwater table and significance of Rabi crops, in all of which the Western region has a clear advantage over the other two. The Bundelkhand region has only a small area under tubewells due to a poor recharge of the crystalline and metamorphic peninsular rocks. The situation in the hills of Dehradun and Nainital districts is also similar.

Decline in Other Wells

In no other field of agriculture, the process of modernisation in Uttar Pradesh is as much evident as in the change from traditional manual or animal worked lifts to tubewells as irrigation method from groundwater sources. Wells irrigated 19.92 lakh hectares of the State's net sown area or 32 per cent of the net irrigated area during the first triennium of the study period; by the last triennium, the percentage came down to a mere 9.62 (9.69 lakh hectares) as well irrigation is not only less efficient when compared to tubewells but even costlier. It now remains important in only few districts where canals and tubewells have not made a headways in Moradabad, Bareilly, Rampur, Badaun and Etah for a complexity of reasons or in Bundelkhand districts in which the groundwater recharge is small. Wells have fallen into disuse or changed into tube wells in most parts of the State.

Stagnant Canal Irrigation

Uttar Pradesh has the country's largest irrigation canal system covering 90.35 lakh hectares. The history of canal irrigation in the State goes back to 1748 when the Eastern Yamuna canal was constructed. It was remodelled in 1830 and now provides perennial irrigation to about 2 lakh hectares in the Upper Ganga-Yamuna Doab. The Upper Ganga canal was built in 1854, when it was the largest irrigation canal in the world. It was irrigates at present about 7 lakh hectares in the Upper and Middle Ganga-Yamuna Doab. The Agra canal, completed in 1873, irrigates 1.38 lakh hectares in the Trans-Yamuna area of Mathura and Agra districts. The Lower Ganga canal (1878) irrigates 5.28 lakh hectares in the Middle and Lower Ganga-Yamuna Doab. Thus the Ganga-Yamuna Doab districts from Saharanpur to Allahabad had a very early start in canal irrigation and this helped the early advancement of agriculture in the region. The Ramganga canal built during the Third Plan has further added to the system; it now supplies water to the Delhi Metropolitan area, augments the capacity of the Upper, Middle and Lower Ganga canals, besides irrigating large area in the Ganga-Gomti Doab. The Project presently irrigates 6.58 lakh hectares.

The Bundelkhand region, which suffered from frequent famines, also received an early attention in canal development, which, however, remained restricted due to the absence of any major river in the area. The Betwa canal (1866) and canals of the Ken, Dhasan and other small rivers now provide protective irrigation to 12.77 lakh hectares in Jhansi, Jalaun, Hamirpur and Banda districts, while the Matatila canal (1966) irrigates about one to two lakh hectares in Jalaun district.

The Central and Eastern regions of the State benefitted as a result of the recommendations of the 1903 Irrigation Commission. The Sarda canal; the largest irrigation system in Uttar Pradesh constructed in 1926, now provides irrigation to about 6.12 lakh hectares. Two other important canals were taken up in the Second and Third Plans : the Sarda Sahayak canal with a potential of 16.5 lakh hectares in the Ganga-Ghaghara Doab and the Gandak canal with a potential of 3.88 lakh hectares in the Trans-Ghaghara plain. A higher rainfall and opposition from the princely State rulers and zamindars (who controlled most of the land in the area till the zamindari abolition in 1956) were the two chief factors responsible for the low development of canal irrigation in this region till recently.

Canal irrigation has been relegated to a secondary position in the State in recent years due to rapid expansion of tubewell irrigation, except in the command areas of the Sarda Sahayak and Gandak and Matatila projects. It now ranks next to tubewells and is used in conjunction with the latter. The decline in its areal importance has not been due to any disuse of canals or reduction in canal irrigated area. In fact, all farmers find canal water more beneficial for crops than groundwater as it replenishes soil fertility and is cost-wise cheaper.

Canal irrigation is also more smooth in operation. It is not only relevant to current irrigation in problems in the State but also holds a high potential and will have to be expanded to reap full benefits of the new farm technology. It was the non-availability of canal water at critical time and in adequate quantity, two very important factors in the HYV technology, which helped in the growth of tubewell irrigation in several localities. But a saturation point has almost been reached in tubewell irrigation in many Western districts and extension of canal irrigation can alone meet the increased demand in such areas as no other method is feasible. A very serious defect in canal irrigation in Uttar Pradesh is that, like in all other regions, it is often used by the farmers in excess which, in turn, leads to the problems of water-logging and saline and alkali efflorescences to prevent which a better method of water management is required. According to the Irrigation Commission Report of 1972, nearly three lakh hectares of land was water-logged and salinity affected 30,000 hectares in the Sarda Sahayak Command Area and 1.95 lakh hectares and 3.52 lakh hectares, respectively in the Gandak Projects Area. Seepage from canals, which is considerably responsible for water-logging and rise in groundwater table can be prevented if the canals are lined. Improvement in natural drainage by constructing drains is essential; and a better management of canal irrigations is also called for looking to its very low efficiency.

Canal irrigated area went up from 22.68 lakh hectares to 30.33 lakh hectares—recording a 33.7 per cent increase, of which 56 per cent was in the Eastern region. The Central region came next with 28 per cent and the Western region followed with 12 per cent. The Hills Region contributed 3.5 per cent and Bundelkhand had only 0.25 per cent. The Eastern region's share in the State's canal irrigated area increased from 13 per cent to nearly 24 per cent, where as the share of the Western region declined from 51 per cent to 41 per cent. Canal irrigation is currently most predominant areally in Bundelkhand, Nainital and Central regions, where it accounts for 70 per cent, 56 per cent and 49 per cent of the net irrigated area. In the Western and Eastern regions on account of faster growth in tubewell irrigation it coves a smaller proportion of 30 per cent and 25 per cent, respectively.

Canal irrigation increased sharply on account of the completion of the Sarda Sahayak canals in Bahraich district (47,000 hectares) and the Gandak canals in Gorakhpur (63923 hectares) and Deoria (1.08 lakh hectares). Also there was a moderate increase in the canal irrigated area in Kanpur (38,664 hectares), Etawah (36,000 hectares), Fatehpur (26,000 hectares), Hardoi (34,462 hectares), Barabanki (38,984 hectares), Varanasi (37,500 hectares) and Mirzapur (38,000 hectares) and in all the other districts of the Eastern and Central regions. Together the increase accounted for 4.16 lakh hectares in the former and 2 lakh hectares in the latter.

Tanks and Other Sources : Only Marginal Significance

Only a small area, about 5.37 per cent (4.86 lakh hectares) of the net irrigated area is currently under tank irrigation and other sources, mostly lifts on diverted streams in all of which there has been a decline in the recent years. Tanks irrigate about 1.96 per cent of the net irrigated area (1.78 lakh hectares) in the State during the last triennium compared to 5.5 per cent (3.4 lakh hectares) in the first, most of it in Basti, Sultanpur, Gonda and Mirzapur, they are of excavated type in the first three and embanked in the last district.

Among other sources, which irrigated 3.7 per cent of the net irrigated area (2.3 lakh hectares) in the beginning and 3.4 per cent (3.08 lakh hectares) towards the end of the study period, *Guls* are important in Dehradun—accounting for 41 per cent of the net irrigated area, and lifts of various types in several other districts of Rohilkhand and in Basti, Gorakhpur and Mirzapur districts. All are very crude, manually operated laborious devices, but all the same quite useful for particularly small farmers who cannot afford other more expensive methods.

Persisting Inter-regional Contrasts

Uttar Pradesh, as stated earlier, is relatively a more advanced State in the field of irrigation. Its net sown area proportion under irrigation is currently about 53 per cent and the gross irrigated proportion in the total cropped are around 46.5 per cent compared to 28 per cent for both in the country as a whole. The growth, however, is very disparate in both and chiefly responsible for the spatially uneven agricultural performance. The Western region has about 38 per cent of the State's gross sown area but nearly half of its gross irrigated area. The two hill districts similarly have a higher per cent of the gross irrigated area (2.03%) compared to the total cropped area (1.39%). In the other three regions, the share of the State's gross irrigated area is smaller compared to the proportion of the total cropped area. The two figures respectively are : 29 per cent and 34 per cent for the Eastern region; 15 per cent and 17 per cent for the Central region; and 4 per cent and 8 per cent for Bundelkhand. The situations in net irrigated area and irrigation intensity are also similar. The analysis clearly indicates a heavy concentration of irrigation development in the Western region and the Nainital district, whereas the other three regions, more so the Central and Bundelkhand, seem to have been neglected. During the study period, though there has been a significant increase in both net and gross irrigated area, the disparity among the regions/districts has mostly remained unaffected, in a few cases it has, in fact, widened. The ratio of net irrigated area between the Western and other three regions was Eastern region 1 : 0.76, Central region 1 : 0.30 and Bundelkhand 1 : 0.01 for the first triennium and 1 : 0.70, 1 : 0.32 and 1 : 0.90, respectively for the last triennium.

Between the first and the last trienniums the State's net irrigated area rose by 47 per cent (from 61,27,756 hectares to 90,35,437 hectares), increasing annually at the rate of 3.13 per cent; gross irrigated area by 56 per cent (from 69,84,761 hectares to 1,09,00,304 hectares), increasing at the rate of 3.56 per cent annum;[1] and irrigation intensity[2] from 114 to 121 per cent. Among the regions, the highest increase (of 103%) in net irrigated area was recorded in the Hills region mainly on acconnt of 118 per cent rise in Tarai area of Nainital district. As a result, nearly 68 per cent of the net sown area is under irrigation in the district and the irrigation intensity is 150 per cent, the second highest after Meerut. Nainital is a lead district in Uttar Pradesh in almost every respect. Besides its large holdings, more progressive farmers and excellent irrigation facilities, Tarai area of Nainital also receives a higher rainfall almost always sufficient for the Kharif crops and thus permitting use of most of the irrigation potential for the Rami crops. The situation in Dheradun its somewhat different, but more satisfactory than in many other districts mainly due to higher rainfall during the year which reduces the dependence upon irrigation to a certain extent.

The Western region had only 49 per cent increase (3.3 per cent per annum) in the net irrigated area and 60 per cent increase (3.73 per cent per annum) in gross irrigated area, but it amounted to nearly half of the total expansion in net irrigated area in the State. Further a decomposition analysis indicated that most of the rise (+111%) was due to tubewells; canals and other sources having a marginally positive

(+6.52 per cent and +6.45 per cent, respectively) contribution, and other wells a significant decline–23%). Tanks' decline also contributed negatively, though, its areas extent was very small. The increase in the net irrigated was over 1,00,000 hectares in Saharanpur, Moradabad and Shahjahanpur; 80,000—97,000 hectares in Bulandshahar, Mathura, Agra, Bijnor, Budaun and Pilibhit and 52,000—78,000 hectares in the remaining ten districts; many of which lie in the Agra division. Net irrigated area now constitutes 60 per cent of the net sown area and gross irrigated area 61 per cent of the total cropped area in this region. Among the districts, as many as 15 have 60—94 per cent of the net sown area under irrigation and the remaining 4 (Bareilly, Bijnor, Budaun and Shahjahanpur) between 50 and 60 per cent. The irrigation intensity is above the State average in all the districts of the region, it being the highest (140-152%) in the four Upper Doab districts of Saharanpur, Muzaffarnagar, Meerut and Bulandshahar with Meerut having the State's highest (152 per cent intensity). The districts below State average include the entire Agra division and Budaun, Moradabad and Farrukhabad in several of which the irrigation proportion is substantially high with a large part of it used for protecting Kharif crops and sugarcane. The trend in irrigation thus indicates a sharp rise in all the districts and mostly due to expansion of tubewell irrigation. In the coming years, a trend towards faster increase in the gross irrigated area in the Upper Doab districts and in both (net and gross irrigated area) in all the remaining districts could be clearly foreseen.

The Eastern region contributed 25.64 per cent to the net irrigated area's and 24 per cent to the gross irrigated area's growth in the State which were respectively 33 per cent (2.3 per cent per annum) and 42 per cent (2.76 per cent per annum). To the rise in the net irrigated area, the contribution of tubewells was about 178 per cent and of canals 61 per cent; other wells' decline had a negative contribution of –11 per cent and tanks and other sources –19 and –7 per cent, respectively. The fall in the other wells' irrigated area was universal and so also was the increase in the tubewell irrigated area and in the net irrigated area itself. Gorakhpur, Deoria and Allahabad districts recorded the highest increase of 1,00,000 hectares, 93,000 hectares and 77,000 hectares, respectively; and Ghazipur, Azamgarh and Gonda districts had over 52,000 hectares each. The irrigation intensity in this region, however, is still low, below the State-average in all the districts, except Varanasi and Mirzapur (134 per cent and 128 per cent, respectively). The regional average is about 122 per cent and the intensity is lower than this in eight districts, *e.g.*, Jaunpur, Ballia, Gorakhpur, Basti, Gonda, Bahraich, Sultanpur and Pratapgarh. The net irrigated area accounts for only 31 per cent of the net sown area in Mirzapur, 38 per cent in Gonda and 19 per cent in Bahraich. Similarly, the irrigation intensity is as low as 101 per cent in Basti and Bahraich and 103 per cent in Ghazipur and Gonda. All these districts are in dire need of improvement. The Sarda Sahayak canals (now under construction) may fill the gap only to a certain extent; a more effective measure will be to provide subsidies and loans for installation of private tubewells which have played a key role in the transformation.

The situation in the Central region inspite of a 69 per cent increase (4.4 per cent per annum) in net irrigated area and 81 per cent (5.6 per cent per annum) rise in the gross irrigated area is also far from satisfactory as only 43 per cent of the net sown area is so far covered with irrigation (49 per cent under canals and 41 per cent under tubewells). Two districts, Sitapur and Kheri, have as low as 33 per cent and 27 per cent coverage, respectively which is very inadequate for the HYV technology and much below the State average; the irrigation intensity currently in these two is 108 (the region's lowest) and 111 per cent, respectively. This region's contribution to the increase in the net irrigated area was only about 15 per cent. Of course, as in the Western and Eastern regions, it was largely due to an expansion of the tubewell irrigated area (82%). The expansion of canal irrigation accounted for 35 per cent of the increase and other wells a negative share of –14 per cent and tanks of –4 per cent. Other sources had a positive share of 1 per cent.

Bundelkhand continues to remain the least developed region not only in Uttar Pradesh but perhaps in the whole country from the point of view of irrigation and, due largely to it, also in agriculture. During the study period, the irrigation increase in this region formed less than one per cent of the total increase and though it was almost entirely (76%) due to expansion of tubewell irrigation, the proportion of net irrigated area fed by this source is still small (5%). Canal irrigation is still predominant (80 per cent of net irrigated area) because of the poor permeability of the peninsular rocks which underlie this region. Bundelkhand is faced with a problem which cannot be solved by the traditional approach. It will require a bold commitment and planning on the part of the government to divert the water from the Ganga (to which reclamation of the ravines is an essential precondition). Only about 22 per cent of the net and gross sown area was under irrigation in this region during the last triennium of the study period. The irrigation intensity is, as a result, also low (101 to 104%). Hamirpur district has the State's lowest proportion (15%) of the net sown area under irrigation and the situation is not better in Banda (19%), Jalaun (26%) and Jhansi (28%). Increase in irrigation facilities is essential to step up yields, to enable farmers to adopt more remunerative cropping patterns and, more importantly to reclaim the degraded wastelands including culturable wastes and fallows that occupy a huge area in the region.

The Rabi crops, which for obvious reasons, take the lion's share of irrigation, accounted for nearly 68 per cent of the gross irrigated area in the first triennium and for 72 per cent during the last. The increase was mainly due to the rise in wheat and sugarcane area for both of which irrigation is a *sine qua non.* The Kharif crops had 24 per cent and 30 per cent, respectively and here the rise was mainly due to a rise in the extent of irrigated area. The said crops similarly increased their proportion from one per cent to three per cent for the two periods due to a higher incidence of irrigation to sugarcane and vegetables and fodder crops which are raised in this season. Crop-wise wheat currently accounts for about 59 per cent of the gross irrigated area and sugarcane and paddy about 10 per cent each, all others being neglected. There is need to step up irrigation incidence to these other crops raised mostly in the Kharif season.

Conclusion

Uttar Pradesh is confronted with problems both due to lack and misuse of irrigation resulting from absence of a perspective planning for the development and management of its water resources. Whereas Bundelkhand and several Eastern and Central districts remain agriculturally poor and backward (due to paucity of irrigation), an unplanned haphazard growth has led to problems of alkalinity and salinity of soils and water logging in several intensively canal irrigated tracts. Besides, drastic fall in the water-table due to indiscriminate sinking of deep tubewells is creating problems of diseconomy for many farmers in the excessively tubewell irrigated tracts of Western Uttar Pradesh.

For a more balanced irrigation development in the plains area of the State, conjunctive use of surface and ground water is essential. This involves stepping up of financial assistance to marginal and small farmers for installation of tubewells, particularly in the Eastern and

Central regions, where a vast majority of farms are below two hectares in size. At the same time, steps should also be taken to improve drainage in the region, particularly in the command areas of the Sarda Sahayak and Ghaghara projects, where water-logging has become an acute problems. A more effective method of water allocation to check over-irrigation from canals and legal measures against indiscriminate sinking of tubewells are equally essential.

The long-term solution to the problem of Bundelkhand lies in transferring Yamuna water and in the construction of embanked tanks in view of the poor acquifer of the peninsular rocks and absence of a major river in the region. This, however, would necessitate reclamation of the degraded lands and a lift-system requiring enormous amount of power, leaving tank construction as a more viable solution at least for the time being.

Uttar Pradesh has enough of water resources to irrigate, upto nine-tenths of its net sown area against only 53 per cent at present. But a careful planning is essential to develop and suitably use these resources.

NOTES

1. Compound growth rate = (Antilog b-1) × 100 obtained by fitting exponential functions of the type $Y = a + bX$ (where Y is the index number and X is the time in log linear form) to the index numbers based on triennium ending 1980-81 for the period.
2. Irrigation Intensity Index= $\dfrac{\text{Gross Irrigated Area}}{\text{Net Irrigated Area}} \times 100$
3. The sum of contributions by all the components equals to the total change, that is, 100. The additive formula used for the component analysis is $I_2-I_1=(T_2-T_1)+(C_2-C_1)+(W_2-W_1)+(O_2-O_1)$ where I represents net irrigated area; T, C, W and O represent, irrigated area by tubewells, canals, wells and other sources, respectively and the figures 1 and 2 denote net irrigated area in the first and last trienniums.

REFERENCES

Dhawan, B. D., 1975 : "Economics of Groundwater Utilisation : Traditional Versus Modern Techniques", *Economic and Political Weekly*, Vol. 10, Nos. 25 and 26, June 21 and 28, pp. A31-42.

Government of India, *Report of the National Commission on Agriculture : Part V : Resource Development*, Ministry of Agriculture and Irrigation, New Delhi, pp. 20-21.

Mellor, John W. and Moorti T. V., 1971 : "Dilemma of State Tubewells", *Economic and Political Weekly,* Vol. 6, March 27, pp. A37-A44.

Moorti, T. V., 1971 : "A Comparative Study of Well Irrigation in Aligarh District, India", *Cornell International Agricultural Development Bulletin,* 19 Cornell University, May.

Moorti, T. V. and Mellor, John W., 1973 : "A Comparative Study of Costs and Benefits of Irrigation from State and Private Tubewells in Uttar Pradesh", *Indian Journal of Agricultural Economics,* Vol. 28, No. 4, pp. 181-89.

Mukhopadhyay, Arunendu, 1973 : "Benefit-Cost Analysis of Alternative Tubewell Irrigation Projects in Nadia District of West Bengal", *Indian Journal of Agricultural Economics,* Vol. 28, No. 4, pp. 189-96.

Patel, S. M. and Patel, K. V., 1969 : *Studies on Economics of Rural Electrification and Lift Irrigation,* Indian Institute of Management Ahmedabad, (Unabridged).

Rao, K. L., 1975 : *India's Water Wealth,* Orient Longman Ltd., New Delhi, p. 103.

Thakur, Jawahar and Kumar, Praduman, 1984 : "A Comparative Study of Economic Efficiency of Different Irrigation Systems in Western Uttar Pradesh", *Indian Journal of Agricultural Economics,* Vol. 39, No. 3, July-September, pp. 521-27.

Under-utilisation of Groundwater Resources, 1980 : "A Case Study of East Uttar Pradesh", *Economic and Political Weekly,* Vol. 15, No. 39, September 29, A113-A122.

17

PERSPECTIVE OF IRRIGATION IN MADHYA PRADESH

S. K. SHARMA AND C. K. JAIN

AMONG THE inputs essential for agricultural development water is one. Its use for irrigation is conditioned by several variables, while low rainfall and its variable nature necessitates the development of artificial means of moisture supply, the increasing employment of fertilisers and, to some extent, high yielding variety of seeds make timely supply of water as a prerequisite. On the other hand, availability of water resource, nature of terrain, and personal, social and economic characteristics of farmers affect the supply side. The present paper is an attempt to analyse the spatio-temporal variation in irrigated area in Madhya Pradesh in perspective of these variables.

It can be hypothesised that the variation in proportion of irrigated area may be related with the variation in the need of irrigation, availability of water resources for irrigation and the capacity of the farmer to use it. The requirement of irrigation depends with other things, on account of rainfall and its variability, nature of crops and the extent of use of modern yield-raising technology. In normal conditions, extent of irrigation should increase with decreasing amount of rainfall and increasing its variability, while it should have direct correlation with the use of modern inputs.

The primary source of water is precipitation, a good deal of which runs-off and is called, 'surface water' and a small amount soaks into the ground and is known as 'ground water'. Thus local availability of water resource is the expression of the rainfall regime. But its feasibility depends on the structure, terrain and soil characteristics. The ground water storage varies with structure and nature of rocks. Further, the nature of terrain goes long way in affecting the practice of irrigation. Level plain as of the Northern Great Plains' favour large scale expansion of irrigation, but hilly and dissected terrain disfavours it. Consequently, amount of irrigation should vary with the nature of terrain. But its actual use depends on the farmer whose capacity to irrigate depends upon his personal perception and knowledge, amount of land he has, his economic capacity, and willingness to invest. It can be presumed that literate and educated farmers may be more prone to opt irrigation. Since economic capacity of farmers depends on the size of holdings, capacity to irrigate should increase with increasing size of holdings. Lastly, perception of the people depends on their social status. People of progressive communities readily come forward to employ modern techniques including irrigation while backward communities take longer time to be convinced. There can be several combinations of these variables, resulting in varigated nature of irrigation. The spatial pattern and growth of irrigated area in Madhya Pradesh has been examined in these perspectives.

Location of the State (between 17°48′ and 26°52′ north latitudes and 74°02′ and 84°25′ east longitudes) in subtropics and concentration of rainfall in few summer months necessitate the creation of irrigation facilities. But the diversified pattern of land and people has manifested in diversified picture of irrigation facilities. Level land is very limited in some of the river valleys such as the Chhattisgarh basin, the Narmada valley. But most of the part is the assemblage of hills, uplands and plateaux. The Malwa plateau in the Western part enclosed by the Madhya Bharat plateau, the Bundelkhand up land, the Vindhyan scarp lands and the Vindhyan range presents a typical character of Deccan trap with undulating surface. The Satpuras, the Baghelkhand plateau and the Bastar plateau are tangled masses of hills and plateau, highly dissected and widely covered with forests. The nature of terrain has manifested itself in the use of water for irrigation and in the levels of agricultural developments.

Spatial Pattern of Irrigated Area (1985-86)

Irrigation is very localised phenomenon in this State. Though, excepting four, all 455 developmental blocks have at least some irrigated area, its proportion varies very widely. In 1985-86, about 30.89 lakh hectares are irrigated out of total 195.2 lakh hectares of total cropped area. Thus gross irrigated area constitutes only 13.9 per cent of the gross cropped area. On block level, irrigated area ranges from nil in Orchha, Kuankondam and Makadi blocks in Bastar and Amarpur block in Mandla, and 0.01 per cent in Gidam and Bastanar blocks of Bastar to 59.0 per cent in Rajim block of Raipur district (Fig. 17.1). It is remarkable that 70.2 per cent (322 in number) of blocks have lower proportion of irrigated area than the average for the State. On the other hand, only one-tenth (49) of blocks irrigate more than one-third of their cropped areas. Out of them, only 13 blocks irrigate more than half of the cropped area.

Blocks with high proportion of irrigated area are confined in three major tracts and four minor areas, besides pockets of isolated block. The irrigated tracts are 1. the Chhattisgarh basin, 2. the Chambal basin, and 3. the Bundelkhand upland. Minor irrigated areas are (i) the Wainganga basin of Balaghat district, (ii) Narsinghpur-Hoshangabad plain, (iii) Indore-Dewas area, and (iv) the Mandsaur area. As would be clear from the analysis of the sources of irrigation, in these tracts and areas canals are major sources of irrigation while they are absent in least irrigated tracts. Therefore, existence and absence of canals has created disparity in proportion of irrigated area.

The heart of the Chhattisgarh basin, comprising blocks of Raipur, Durg and Bilaspur district is the first ranking irrigated area in the state. It is also the first region to get irrigation facilities. This 'rice-bowl' has rainfall more than 120cm which is highly variable causing frequent droughts and famines. The famines of 1896 and 1899 ushered in a period of tank building as protectional measures against recurring these famines, resulting in growth of irrigation facilities before 1950-51. Actually the interflue area of the Mahanadi and Seonath is the major area in the Chhattisgarh. This is the designed area of the Tandula and the Mahanadi canals. Other worth mentioning projects are Khapri, Kumhari, Pindrawan and Dudhwa in this area. The Hasdo basin in Bilaspur is another significant irrigated sub-area of this region. Kharun and Maniyari are other important projects in this sub-area. Thus, the agriculture of the major portion of the region is least affected by the development of irrigation facilities localised in the core area.

The second highly irrigated tract extends over the Chambal basin of Gwalior, Morena, Bhind, Shivpuri and Datia. Most of the blocks have more than 30.0 per cent of cropped area under irrigation; three of them, viz. Kailaras, Bhitarwara and Sheopur have irrigated more than half of the cropped area. This is the region of lowest rainfall and highest rainfall variability in the state. It has good fertile alluvial soils and comparatively level tracts. In this area also, several medium and minor

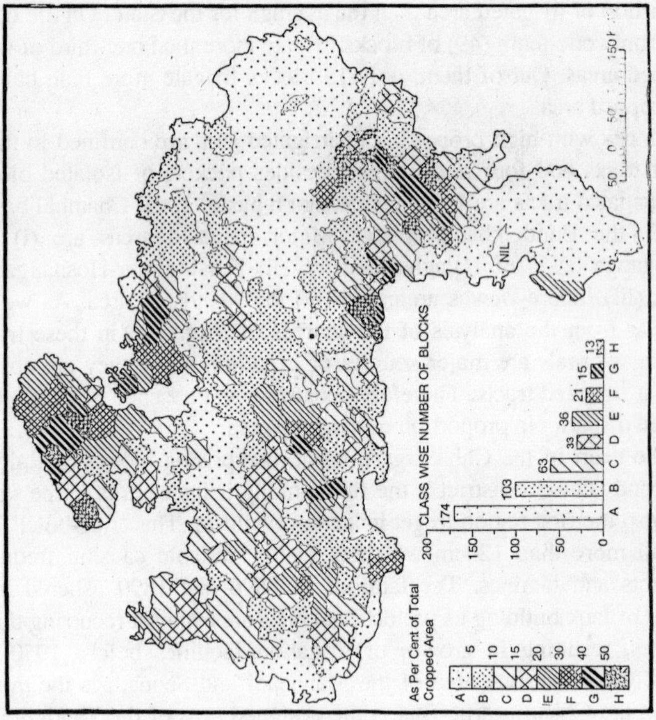

FIG. 17.1: Madhya Pradesh: irrigated area 1985-86.

projects and a few major projects have been completed. Among them, the Chambal projects, Harsi, Bhander, Awadha, Sindh, Tigra and Pagara are worth mentioning. It is also predominantly canal irrigated tract.

The Bundelkhand upland, comprising the blocks of Tikamgarh and Chhatarpur districts, is the third major irrigated tract. In fact, invening blocks of Lalitpur and Jhansi districts of U.P. between Tikamgarh and Shivpuri, and Datia also have high proportion of irrigated area. Thus, barring the state boundary, the Bundelkhand-Chambal basin make a continuous tract of high irrigated area. It differs from previous tracts in respect of means of irrigation. Exceptionally, this area depends on wells and tubewells while the Chambal area depends on canal systems. Gneisses and granites of this popularly known as the Bundelkhand gneisses, do not have rich acquifer, however, fractures and joints in them provide sufficient recharge in wells. Water table is comparatively high. Such conditions favour the construction of wells and tubewells.

The Wainganga valley of Balaghat district is one of the secondary irrigated tract, where three blocks, viz. Katangi, Lalbarra and Waraseoni have more than half of cropped area under irrigation. The Wainganga and Bagh projects provide water for irrigating more than 46 and 15 thousand hectares respectively. Located in high rainfall zone, the rice crop of this area required regular and ensured water supply. After the completion of the Tawa project on the same river, a tributary of the Narmada river, Hoshangabad district has emerged as a irrigating tract. On average, 25.5 per cent of the cropped area of the district is irrigated. However, Timarni, Harda and Seoni-Malwa in western part and Babai, Sohagpur and Piparia in the east have proportion of irrigated area more than one-fifth of the cropped area. Three contriguous blocks (Udaipura, Silwani and Bareli) of Raisen also have comparatively higher proportion of irrigated area. The Barma project on the same river is the major source of irrigation in this area. Its offshoots is extending in Sehore and Bhopal districts in the west.

Proportion of irrigated area more than the state average is in thirteen blocks around Indore. It ranges between 15.0 to 34.0 per cent of total cropped area in these blocks. Three of them (Indore, Mhow and Sawer) are in Indore, three (Thikari, Maheshwar and Barwah) in Khargone; three (Dharampuri, Umarban and Manawar) in southeastern Dhar; three (Dewas, Sonkachh and Bagli) in western Dewas and only one (Ujjain) in Ujjain district. It is worthmentioning that it is predominantly a well and tubewell irrigated tract, though there are several minor irrigation projects here. Last significant secondary irrigated area is constituted by six blocks of Mandsaur district in the northwest. Though such big reservoir as the Gandhisagar on the Chambal is in this district, this area solely depends on wells, not on tubewells, for irrigation. Resultantly, irrigated area range between 15.0 and 26.0 per cent of the cropped area, which is between 100 and 200 per cent of the state average.

Conversely, vast area of the state is still lagging far behind in this basic facility. It is as low as less than 1.00 per cent of cropped area in 48 blocks (10.5%). Thus, they are extremely poor. Another 126 (27.5%) of blocks irrigate between 1.0 to 5.0 per cent of cropped area and therefore belong to very low class. 103 blocks (22.4 per cent of total) belong to low class, irrigating between 5 and 10 per cent of cropped area. Only 45 blocks (9.9 per cent of total) belong to lower medium class.

Distribution of these blocks show that southern and eastern half of the state except the core of Chhattisgarh, constitute the least irrigated part in the state. In Bastar (1.9 per cent of cropped area), Surguja (3.0), Sidhi, Shahdol (2.2) and Mandla (2.3) districts irrigation has not started yet in real sense. In western M.P. none is so poor. The Bastar plateau, Baghelkhand plateau, Satpura region, Rewa plateau, northwestern and southeastern Malwa and the fringe areas of the Chhattisgarh basin are poor to extremely poor in irrigation facilities. Because of this fact, yield rates of crops, consequently, productivity of agriculture is low to very low in these regions. Since majority of development blocks belong to classes below the average of the state, it is imperative to analyse the perspectives of irrigation and obstacles in the way of rapid expansion of irrigation facilities.

Growth of Irrigation

Irrigation has been practised since quite a long time in a few parts of the state. The Gwalior region in the northwest and the Chhattisgarh region in east are worth mentioning, where irrigation potentials were created during 1920-40 (Sharma and Jain 1987, 44). But in other parts, development of irrigation is the creation of the plans. In the beginning of the first Five Year Plan (1950-51) gross irrigated area was 905 thousand ha. which was only 5.9 per cent of gross cropped area. During last 35 years of plan regime 7 major, 101 medium and 4965 minor projects have been executed and completed. As a result of these efforts total irrigated area increased to 3106 thousand hectares in 1984-85, constituting 13.9 per cent of the cropped area and recording an increase of 248 per cent. The progress of irrigated area is furnished in Table 17.1.

TABLE 17.1: M.P. : Progress of irrigation 1950-85 (A = Area in thousand ha)

Year	Canals		Wells		Tanks		Other		Total	
	A	%	A	%	A	%	A	%	Net A	Gross irrigated A
1950-51	354	39.8	249	28.0	235	26.4	52	5.8	890	905
1955-56	370	4.8	298	36.0	119	14.4	38	4.8	825	837
1960-61	441	7.7	324	35.1	122	13.2	37	4.0	924	938
1965-66	461	47.2	345	35.3	118	12.1	52	5.4	976	1000
1970-71	710	48.0	551	37.2	130	8.8	90	6.0	1481	1523
1975-76	801	44.4	700	38.8	132	7.3	171	9.5	1804	1896
1980-81	1035	44.4	986	42.3	136	5.8	175	7.5	2332	2453
1984-85	1267	42.1	1285	42.7	204	6.8	254	8.4	3010	3106

Source: Directorate of Agriculture, M.P. Agricultural Statistics, 1985.

It took about fifteen years to add 95 thousand hectares to the irrigated area from 1950-51 to 1965-66, but in next decade another 896 thousand hectares were brought under irrigation. By the end of the third plan, several irrigation projects had started irrigation. Consequently, irrigation potentials only from government source increased from 158 thousand hectares to 610 thousand hectares of major, from 195 thousand hectares to 483 thousand hectares of medium and from 111 thousand hectares to 834 thousand hectares of minor projects, from pre-plan period to 1987-88. Of the total created potentials of 1927 thousand hectares only 73.5 per cent (1417 thousand hectares) could be utilised for irrigation.

There is a wide range of variation in growth rate of irrigated area during 1950-85. Total irrigated area of the state increased by 243.0 per cent, while the mean of the district rates is very high (i.e. 1036). The very high standard deviation (1461) of these rates also proves very high variability. Growth rate varies from –32.6 per cent in Bastar to 60.39 per cent in Vidisha district (Fig. 17.2). Very high growth occurred where initial proportion of irrigated area was extremely low. In 31 out of 43 districts, growth rates, were higher than the state average and ten of them Rewa, Sidhi, Shahdol, Surguja, Mandla, Satna, Narsinghpur, Hoshangabad, Vidisha and Raisen had less than one per cent, eight (Sagar, Damoh, Panna, Khandwa, Jhabua, Sehore, Dewas and Ujjain) between one and two per cent and six (Guna, Indore, Dhar, Khargone, Jabalpur and Chhindwara) between 2 to 3 per cent of total cropped area under irrigation in 1950-51. Vidisha recording highest increase of 6039 per cent, had only 0.17 of cropped area under irrigation. Only seven districts had irrigated area more than 3.0 per cent. Resultantly, coefficient of correlation between initial proportion and rate of growth is negative and moderately significant (r = –0.43). The phenomenal increase in least irrigated district is the result of very small denominator. Even at present, thirteen districts have low to very low proportion of irrigated area (Table 17.2).

TABLE 17.2: M.P. : Districts according to present status and rate of growth of irrigated area during 1950-85

Present Status	High	Medium	Low
Highly Irrigated (12)	Morena, Datia, Indore, Hoshangabad	—	Bhind, Gwalior, Shivpuri, Raipur, Bilaspur, Balaghat, Tikamgarh and Chhatarpur.
Medium Irrigated (20)	Shajapur, Dewas, Sehore, @, Raisen Narsinghpur, Dhar, Khargone, Khandwa and Ujjain.	Betul, Rajgarh, Panna, Jabalpur, Chhindwara, Mandsaur and Ratlam.	Durg * and Seoni.
Low Irrigated (13)	Vidisha, Mandla, Sagar, Rewa, Sidhi, Shahdol, Satna & Jhabua.	Guna, Damoh, and Surguja	Raigarh and Bastar (declined).

@ Sehore includes Bhopal and * Durg includes Rajnandgaon also.

And thus, quantum of denominator has been major determinator of growth rate of these districts.

The districts with low and very low rates of growth also present inverse relationship. There are eleven such districts, ten of them had high and very high proportion of area under irrigation. It increased at very slow rate (below 50%) in three districts (Bilaspur, Raigarh, and Balaghat) and at slow rate in others (Raipur, Durg, Seoni, Shivpuri, Gwalior, Bhind, Tikamgarh and Chhatarpur). In Bastar, absolute as well as percentage irrigated area declined drastically. It can well be remarked that districts with initially poor irrigational base are still poor, despite a high and very high percentage of growth. Problem is more acute where growth is also slower with poor base. These districts require special case in planning of agricultural development.

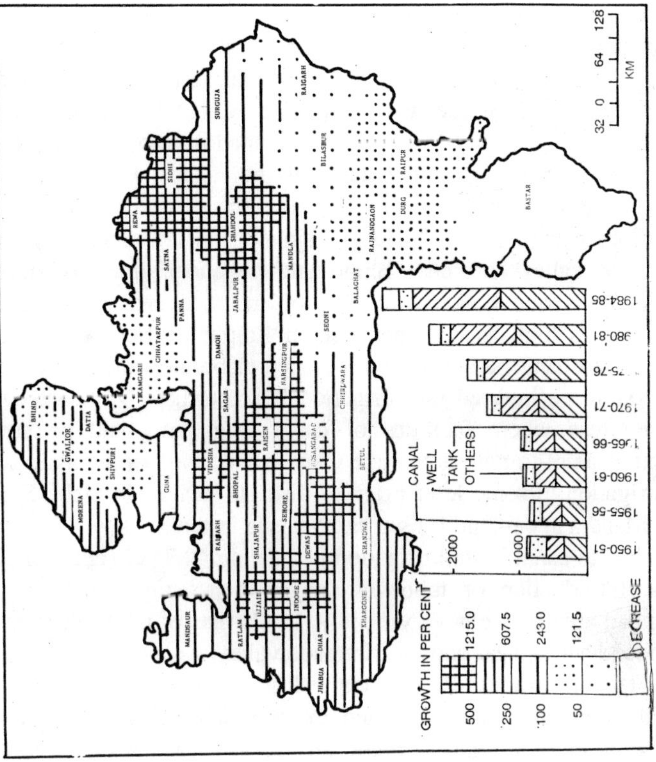

FIG. 17.2: Madhya Pradesh: changes in irrigated area 1950-51 to 1984-85.

Means of Irrigation

Irrigation is done with the help of canals or by lifting sub-soil water. The physical conditions have caused spatial variation in utilisation of surface and sub-surface water resources. Canal is the major source providing water for 42.1 per cent of irrigated area, but it is large scale adventure and is out of the reach of individuals. Further, it cannot reach very point. It is almost single significant source of irrigation in the Chhattisgarh (Raipur, Durg, Rajnandgaon and Balaghat) and in Chambal valley (Morena, Bhind and Gwalior). Intensity of canal irrigation is above 60.0 per cent of irrigated area in these areas reaching as high as 82.8 per cent in Raipur district. As mentioned earlier, Hoshangabad also has developed elaborate system of canals. It can safely be inferred that irrigation is somewhat significant only in those areas where facilities of canal irrigation could be created. Another point to note is that the canal is emerging as the main source of irrigation in some least irrigated districts such as in Sidhi, Shahdol and Mandla in east and Raisen and Vidisha in centre.

Next to canal is well irrigation inundating 42.7 per cent of irrigated area. Introduction of tube-well irrigation has raised proportion of irrigated area by wells very rapidly. In districts of the Malwa plateau, Bundelkhand upland and the Narmada valley, well is the major source of irrigation. In all in 22 districts well is the first ranking source. It is worth mentioning that only ten districts with higher proportion of irrigated area than mean depend on sub-soil water resource lifted by several means. These are Tikamgarh, Chhatarpur, Shivpuri, Datia, Indore, Ujjain, Dewas, Mandsaur, Shajapur and Khargone. Along with 1034 thousand dug wells used for irrigation purposes, 11374 tube-wells have been commissioned upto 1984-85. They are concentrated in the Malwa, Narmada valley, Chambal basin and the Chhattisgarh basin. In fact, in Western Malwa is basically dependent on tube well irrigation and this excessive drafting of underground water ignoring the surface water has resulted in lowering the water table and drying of dug-wells particularly in summers (Sharma 1987 p. 247).

Tank irrigation, which had been as significant as well irrigation in pre-plan period, declined absolutely and relatively. At present it provides water for only 6.8 per cent of net irrigated area. It is only significant in the Chhattisgarh region, highest intensity being 55.3 per cent in Bastar. Outside this region Balaghat, Seoni and Khargone are significant. There are more than 55 thousand tanks of various sizes utilised for irrigation in the state out of which 42 thousand are in Raipur and Bilaspur divisions, and more than 7 thousand in only Balaghat district. Predominance of tanks in this area may be ascribed to several physical facts such as undulating terrain with impervious rock-base and high rainfall. Water is utilised in rainy season itself for rice, the porosity of soils does not prove hazardous. It is to be noted that catching of fish is also done in these tanks.

Other sources such as nallah anicuts etc. irrigated 254 thousand hectares (8.4%) in 1984-85. It is found that least irrigated districts invariably have higher proportion than the average and all districts with higher proportion irrigated by these sources have lower percentage irrigated area than mean.

Rainfall and Irrigation

Rainfall can be one of the determinants of both the requirement of irrigation and availability of water resources. Virtually, irrigation is a means to meet the deficiency of rainfall. Therefore, if other things remain constant, need of irrigation should increase with the decreasing amount of rainfall, presenting inverse relationship. Thus there is greater probability of development of irrigation in low rainfall zones entirely. But in the case of Madhya Pradesh it is not true. Areas of concentration of irrigation are not necessility areas of low rainfall. Average annual rainfall in the state ranges from 212 cm at Pachmarhi and 179 cm at Bhanupratappur in Bastar to 62 cm at Gohad in Bhind in the northwest and 78 cm at Jhabua in the west. The southern and eastern parts receive heavy rainfall which declines gradually towards north and west. Rainfall is lower than 130 cm north and west of the Satpura range while it is higher south and east of it. Major portion of the Malwa plateau receives rainfall between 80 to 100 cm and the Chambal basin below 80 cm per annum. If this pattern rainfall is compared with the pattern of irrigated area its growth in the Chambal basin and in the Bundelkhand upland can safely be ascribed to the low rainfall; sub-areas of the Hoshangabad plain and the Indore area are also developing due to low rainfall. Almost entire Malwa region, right from Raisen in the east to Indore in the west, is picking up irrigation to compensate this deficiency. Contrary to them, Chhattisgarh and the Wainganga basin of Balaghat receive comparatively high rainfall. Nevertheless, they constitute areas of concentration of irrigation, which may be due to reasons other than the amount of rainfall.

Rainfall variability is another aspect influencing the distribution of irrigated area along with the amount of rainfall. The coefficient of variation in rainfall, calculated utilising the annual rainfall data of seventeen years between 1956-57 and 1973-74, ranges from only 10.85 per cent in Shahdol district to 48.93 per cent in Khargone. The distributional pattern of rainfall variability does not present apparent inverse relationship with the amount of rainfall. However, it is comparatively low in the high rainfall areas of south and east and high in low rainfall areas of west and north. Coefficient of rainfall variation is comparatively low (below 26.0%) in the Chhattisgarh region; inspite of this proportion of irrigated area is high. As stated earlier, the root of the genesis of this facility in this region is in recurring droughts and famines. Similarly, rainfall is less variable (below 24.0%) over the Bundelkhand upland of Tikamgarh and Chhatarpur, and also over the Chambal basin (less than 30%) excepting Bhind where it is very high, i.e. 48.4 per cent. Thus, it is the low amount of rainfall which necessitated the development of irrigation in these areas.

Contrary to them, rainfall is low and highly variable (above 30%) in the Malwa region and in the Middle Narmada basin, but irrigation could not be developed commensurately. Eight districts (Dhar, Jhabua, Ujjain, Shajapur, Dewas, Khargone, Khandwa and Betul) of this part have been identified as drought prone districts also (S. Balkrishna et al 1982 pp 58-67). Under the drought prone area programme some of these districts progressed well in developing irrigation facilities and proportion of net irrigated area reached to 6.4 per cent in Jhabua and 11.4 per cent in Khandwa to 14.5 per cent in Dewas among these districts in 1984-85. There are three other districts classed as drought prone. Datia is one of them which has increased the percentage of irrigated area from only 3.1 per cent in 1950-51 to 21.8 per cent in 1981-85. But other two districts, viz. Shahdol and Sidhi are still very poor in this respect with only 2.2 and 3.7 per cent of cropped area under irrigation respectively. Thus, the amount and variability of rainfall have been one of the major considerations behind the development of irrigation facilities, but they have not been sole basis and that's why certain areas fulfilling these qualifications are still deprived of this facility.

Land Capability and Irrigation

Investment in irrigation also depends on facets of land, called land capability. If land is capable of yielding comparatively high return, there would be higher chances of employing capital intensive inputs on them. Even such basic input as irrigation requires comparatively level land covered with fertile soils for operation and economic reasons. It is because of this fact, major and minor irrigation tracts of the state are confined on very good to moderately good lands. The land of the Chhattisgarh basin covered with red and yellow soils is rated very good and good in the land capability classification (S.K. Sharma, 1988, a, p 127). Old alluvial and mixed red and black soils on level land of the Chambal basin and part of the Bundelkhand upland make them agriculturally productive areas. These characteristics of land resource facilitated the use of water for irrigation in these areas. These are agriculturally high productive areas in the state (Sharma and Jain, 1985, p 36). Malwa and the Narmada valley also possess good and moderately good lands. The Wainganga basin and the Mandsaur area also have good quality land. However, despite extensive good land over the Rewa plateau, facilities could not be created on large scale. At the sametime, northeastern Malwa and the Jabalpur haveli could not pick up irrigation in spite of the presence of good and moderate good lands.

Socio-economic Perspectives and Irrigation

Decision of irrigation is taken in the context of the existing socioeconomic conditions of the society. It has been true in the case adoption of other agricultural inputs (Sharma, 1988 b; Sharma and Jain 1985 pp 51-53). These characteristics range from the personal characteristics of peasants (Ilbery, 1978 p 448), to such social structure as ownership and distribution of land and cultural land technological status of the society. With this perspective in mind, three factors have been selected for discussion. These are literacy among cultivators, proportion of tribal population and the distribution of operational holdings by size.

Literacy among Cultivators

Literacy can be taken as an index of awareness. Proportion of literate cultivators to total in rural areas ranges from 4.4 per cent in Jhabua and 11.3 per cent in Bastar to 48.0 per cent in Narsinghpur and 48.2 per cent in Indore in 1981, with the state average of 13.7 per cent. Literacy among cultivators has not close and significant relationship not only with the extent of irrigated area but with other yield-raising inputs. For instance, highest literacy among cultivators was in Indore but it is at 11th place in proportion of irrigated area (i.e 21.0%) and at 19th place in the user of HYVs of seeds. Jhabua with lowest literacy among cultivators is not at the 36th place among 45 districts of the state. It results in very low coefficient of correlation (r = +0.10) between the proportion of literate cultivators and the extent of irrigated area. It is worthmentioning that proportion of literate farmers is below one-fifth in the Bundelkhand upland and below one-third in the Chambal basin and Chhattisgarh basin. In fact, there is little chance of choice in major irrigating tracts, because major sources of irrigation are governmental canals and tubewells and farmers of the command areas have to irrigate. On the other hand, the Rewa plateau and Jabalpur haveli, with comparatively high proportion of literate farmers, could not come forward to develop irrigation.

Scheduled Tribes and Irrigation

In comparison to literacy proportion of scheduled tribes in total population has deeper bearing on practice of irrigation. Because of the limitations put forth by the land and more because of their own socioeconomic limitations these people do not practise irrigation. About 12 million people are tribal in the state (1981) who are 23.0 per cent of total population. There are 15 districts out of 45 with proportion of tribal population higher than the state average. Highest proportion is in Jhabua (83.5%) in the west followed by Bastar (67.7%) in south and Mandla (60.4%) in the central part of the state. Tribal population constitutes more than half of total population in Surguja and Dhar and more than two-fifths in Raigarh, Shahdol and Khargone districts. In Seoni, Betul, Chhindwara and Sidhi more than 30.0 per cent and in Khandwa, Bilaspur and Rajnandgaon less than 30.0 per cent (but higher than the state average) of population is scheduled tribe. On the basis of proportion of tribal population 176 development blocks out of 459 have been recognised as tribal blocks. Their distribution is in five zones which extend over the Bastar plateau, Baghelkhand plateau, Raigarh hills, Satpura ranges and plateaux and Dhar uplands. The land of these tribal areas is hilly and dissected and possesses low inherent fertility; and therefore physically is not capable of cultivation. Though heavy stress has been put on the agricultural development in these tribal blocks by providing facilities for adopting modern technology, they remain far behind in this respect (S.K. Sharma 1989, 71-80). It is accepted that tribals have their own world and live in their own isolated ecosystem and the logic of adoption of yield-raising technology is beyond their level of comprehension. Consequently, irrigated area is extremely low in these areas. The average of irrigated area for tribal blocks is hardly 4.4 per cent of the cropped area. More than two-thirds (119) of tribal blocks have less than 4.0 per cent of their cropped area under irrigation in 1985-86. This proportion is less than 30.0 per cent of the state average. Out of them, 55 blocks have as low as only one-tenth of state average. Contrary to it, only ten blocks have higher proportion of irrigated area than this average, while the density of pumping sets is higher than the state average in 35 blocks. Out of them, six are confined in the western zone, two in south-central zone and only one in the southern zone. Contrary to it, more than 80 per cent blocks of southern zone, 84.4 per cent of eastern zone and 100 per cent of northeastern zone have proportion of irrigated area less than 30 per cent of the state average. Not only that irrigated area dwindled in Bastar district. Thus, tribal areas are still to come up on the irrigation map of the state.

This low level has been ascribed to the weak economic base of the people and, therefore, provisions for granting subsidies were made. Such incentives could not make exogenous method and techniques attractive to them. On the other hand, non-tribal people of the area utilised these facilities on the name of tribals. Even tribals have made it a fashion to obtain irrigation pumps on subsidy and sell them out to non-tribal farmers on lower cost than existing in the market. In this manner, social status plays decisive role in decision-making.

Size of Holdings and Irrigation

Like manufacturing units, size and ownership of operational holdings in agriculture determine the use of inputs and efficiency of farms. Several yield-raising technologies along with irrigation mechanism have been made available but they are not neutral to size, and hence they are not within reach of every peasant. Being capital intensive the use of these technologies and inputs depends, with other things, on the capacity of farmers. In turn, this capacity is determined by the size of holding. Nevertheless, small and marginal farmers, who possess limited land, always try to utilise it intensively. To maximise output and to intensify the use of land these farmers use such inputs which are within their reach. Due to financial constraints they cannot go to capital intensive inputs and technologies. Big and well-to-do farmers, on the other hand, are capable of spending on capital intensive inputs but they have comparatively lesser interest in intensification of cultivation, because there is no scarcity of land for them, compelling to intensify the landuse. It is true in the case of irrigation also.

On average, 28.3 per cent of holdings were wholly or partially irrigated and remaining 71.7 per cent were wholly unirrigated in the state in 1980-81. These proportions vary enormously with the size (Table 17.3). Irrigation is practised by comparatively much smaller proportion of marginal (below 1 ha), small (1-2 ha) and semi-medium (2-4 ha) land holders in comparison to big farmers. More than 46.0 per cent of holdings with size above 20 ha practise irrigation, while proportion of irrigating farms of marginal class is only half of this, i.e. 23.0 per cent and of small class is only one-fifth (25.6%). Thus only a few marginal and small farmers can utilise the facility of irrigation. It presents the inverse relationship between the size of farms and proportion of holding practising irrigation. However, proportion of net irrigated area to net sown area presents inverse relation with the size of holdings. On average 12.5 per cent of net sown area as net irrigated in 1980-81, while proportion is higher than this for holdings below 5 ha in size, highest being 18.3 per cent for holdings below 0.5 ha. It is true for least irrigated district like Bastar, highly irrigated district such as Tikamgarh and for the district predominated by marginal, small and semi-small holdings, like Bilaspur (Table 17.3). It can be expected that proportion of irrigated area should be higher in districts which are predominated by smaller sizes of holdings. This is proved by moderately positive coefficient of correlation (r = +0.41) between proportion of holdings under 4 ha and percentage of net irrigated area of forty five districts of this state.

Significance of sources of irrigation also vary with the size of holdings. Marginal and small farmers depend on public means of irrigation because they hardly can afford for their own sources. But big landlords believe in having their own sources along with the public means of irrigation. As such, between 55.0 to 64.0 per cent of net irrigated area of Marginal and small farms get water from canal, while on larger farms canal irrigated area constitutes 34.0 to 50.0 per cent of net irrigated area (Table 17.4).

TABLE 17.3: M.P.: Proportion of irrigated area and irrigating holdings by class-size, 1980-81

Size class	Net irrigated area as % of NSA					% of irrigating holding to total				
	M.P.	Bilaspur	Bastar	Tikamgarh	Hoshangabad	M.P.	Bilaspur	Bastar	Tikamgarh	Hoshangabad
Below 0.5	18.3	12.8	1.5	42.0	17.2	21.7	26.8	3.3	55.8	13.1
0.5-1.0	17.8	20.3	1.3	42.0	22.1	24.8	27.3	3.8	58.3	28.0
1.0-2.0	15.5	19.9	1.3	36.7	21.1	25.6	27.9	4.6	58.8	28.8
2.0-3.0	14.2	19.5	1.4	40.2	19.6	27.2	27.0	4.9	59.0	28.4
3.0-4.0	14.6	19.5	1.5	46.1	19.9	32.0	29.6	7.4	71.3	33.7
4.0-5.0	13.8	16.6	2.1	44.7	17.3	33.0	27.0	8.5	73.1	32.7
5.0-7.5	12.2	17.8	1.6	38.8	16.1	34.6	27.7	9.8	74.8	35.8
7.5-10.0	12.0	15.9	1.4	43.7	15.5	38.0	30.4	8.7	79.0	36.8
10.0-20.0	9.8	14.1	0.9	40.5	12.4	39.3	30.2	7.1	78.0	38.8
20.0-30.0	9.8	17.5	0.3	48.1	14.0	45.2	39.7	4.9	85.8	43.2
30.0-40.0	9.8	12.4	0.3	38.3	16.1	46.9	35.8	4.0	73.0	47.1
40.0-50.0	10.0	14.6	0.1	20.9	15.1	50.5	41.5	2.3	60.0	52.6
Above 50	11.6	21.0	0.2	48.4	18.9	49.4	54.9	11.1	63.6	61.8
All	12.5	18.1	1.1	41.1	15.7	28.3	39.2	5.7	61.8	31.2

Source: Agricultural Census of Madhya Pradesh, 1980-81.

TABLE 17.4: M.P.: Percentage of area irrigated by different sources by size of holdings, 1980-81

Size	Canal	Tank	Well	Tubewell	Others	All
below 0.5	63.6	8.0	22.5	1.6	4.4	100.0
0.5-1.0	60.0	6.8	26.5	2.3	4.4	100.0
1.0-2.0	55.3	5.9	32.0	2.6	4.2	100.0
2.0-3.0	50.3	5.8	36.6	2.3	5.0	100.0
3.0-4.0	47.9	5.5	37.8	3.0	5.7	100.0
4.0-5.0	44.7	5.4	39.8	3.4	6.6	100.0
5.0-7.5	42.2	5.3	42.1	3.1	7.3	100.0
7.5-10.0	37.5	7.1	43.4	3.2	8.8	100.0
10.0-20.0	34.4	5.5	47.8	2.6	9.7	100.0
20.0-30.0	36.3	6.0	43.7	3.5	10.6	100.0
30.0-40.0	38.0	5.0	40.1	3.8	13.0	100.0
40.0-50.0	42.8	4.0	37.0	3.4	12.8	100.0
Above 50.0	49.6	6.2	23.1	4.5	16.6	100.0
All	44.4	5.8	39.4	2.9	7.4	100.0

Source: Agricultural Census 1980-81, Report pt. 2.

Further, well is the second source of irrigation for small as well as big farmers. This is to note that farmers with smaller holding than 4.0 ha are getting subsidy for digging wells and purchasing pumping sets. Consequently density of all pumping sets per 100 ha is very high on marginal (216) and small (159) holdings, but comparatively much lower (only 17) on farms higher than 10.0 hectares. It is pertinent to note that percentage irrigated area by tubewells ranges from 1.6 per cent in case of holdings below 0.5 ha to 4.5 per cent in the case of holdings larger than 50.0 hectares. In fact, most of tube-wells are in public sector which benefit all farmers within their reach, and therefore there is not much variation by size.

Consciousness of the People

In the present political set-up of the country decisions go by the consciousness of the people. It has been elaborated elsewhere (Sharma 1988 a p. 141) — that the developmental efforts are influenced by the consciousness and awareness of the people. In the context of the country this state is regarded as backward state. Functionally it seems true also. As a consequence, rivers, flowing almost in all directions, drain huge water resource and as soon as they cross the boundary of Madhya Pradesh they are deemed in such a fashion that the submerged area confines in this state but irrigated area in other states. This is true in case of all boarding states and all streams and rivers. It is suggestive that this state is unable to perceive the significance of water resource.

Intra-state variation also seems to be closely related with the consciousness of the people. As elaborated, tracts of low conscious tribal people stand as least irrigated. Even in other areas extent of the irrigated area is related with awareness of the people. As an index of consciousness or awareness of people, the voting behaviour can be taken. The percentage of voters actually turned up to total voters in Parliamentary election 1989 in each legislative assembly constituency of this state has been plotted on (Fig. 17.3). Because of differences in number and territorial extent, correlation between this and extent of irrigated area cannot be calculated, however, it shows very close resemblance with the distribution of irrigated area. It supports the notion that use of this resource is subjective and restricted.

Irrigation and Adoption of Yield-raising Inputs

Besides irrigation, there are other inputs used for raising productivity. Among them, fertilisers and high yielding varieties of seeds have been important. The use of these inputs also increased phenomenally, after second five year plan. All there were important components of the 'Green Revolution'. It is therefore expected that use of one may affect the use of others.

Irrigation and Area under HYVs

It is fact that the benefit of irrigation multiplies with the use of improved seeds and it is also true that the water requirements of most of the improved varieties is much higher than the local varieties (Sinha, 1975). As a consequence of the green revolution programme, several dwarfish varieties of seeds were evolved and introduced. By the end of 1966-67 area under such seeds was 57.4 thousand hectares, which was only 0.48 per cent of area under all cereals. It increased to 4409.3 thousand hectares in 1984-85, constituting 33.9 per cent of cereal cropped area. There is close comparability in spatial distribution of these two phenomena. There are five tracts of the concentration of the use of HYVs. They are the Chambal basin, Bundelkhand upland, Chhattisgarh basin, Hoshangabad plain and adjoining Satpura region and the western Malwa region. In all, 21 districts have higher proportion of cropped under HYVs; all of them, excepting four, have proportion of irrigated area above the state average, i.e. 13.9 per cent, Consequently, coefficient of correlation between proportion of irrigated area and percentage area under HYV is positive and highly significant ($r = +0.76$).

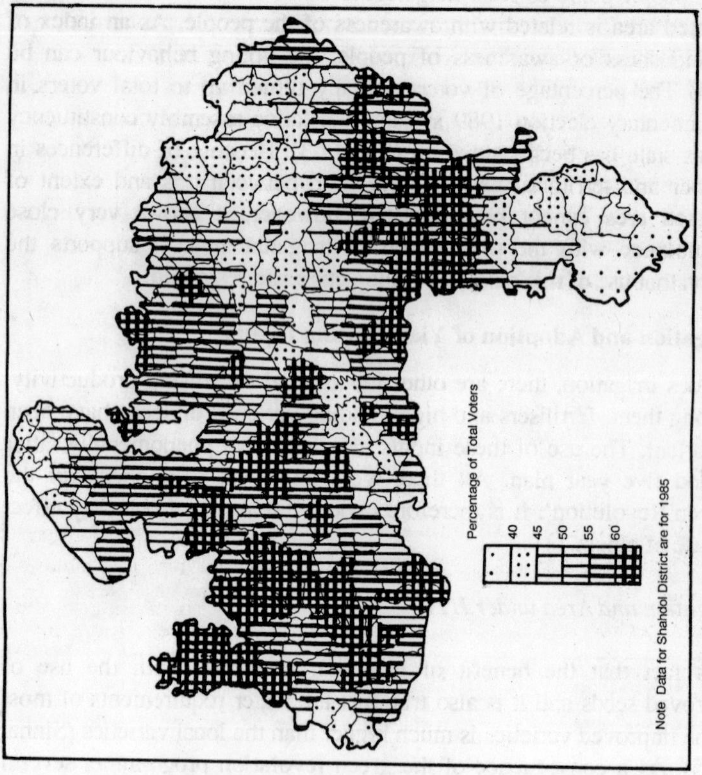

FIG. 17.3: Madhya Pradesh: proportion of voters Lok Sabha election 1989.

Irrigation and Use of Fertilisers

Use of fertilisers has also assumed significant place. On average, 16.6 kgs. of chemical fertilisers were used per hectare of cropped area in 1984-85 in this state which is a little more than one-third of the average (42.8 kgs/ha) for the country. Poverty of farmers, lack of ensured and timely water supply and subsistence nature of cropping may be the reasons of this state of affairs. However, distribution of fertilisers has recorded an increase of more than 125 times, from only 2.96 thousand tonnes in 1956-57 to 372.6 thousand tonnes in 1984-85. Annual average of use of fertilisers per hectare increased from 0.17 kg. to 16.6 kgs. during this period. This growth is very similar to thatoof irrigated area.

There is wide variation in spatial patterns of use of fertilisers. Per hectare use ranges from only 2.4 kgs. in Mandla district to 56.7 kgs. in Morena district. There are four areas of concentration viz. the Chambal basin, (Morena, Bhind and Gwalior) core of the Chhattisgarh basin (Raipur, Durg and Bilaspur), the Narmada valley (Jabalpur, Narsinghpur, Hoshangabad, Khandwa and Khargone) and the major portion of the Malwa plateau (Dhar, Ratlam, Ujjain, Mandsaur, Indore, Dewas, Shajapur, Sehore, Raisen and Vidisha). Out of these areas, Tikamgarh, Balaghat, Rewa and Satna districts also have high use of fertilisers per hectare. These districts have comparatively high and medium proportion of irrigated area. It yields to highly positive correlation ($r = +0.73$) between proportion of irrigated area and use of chemical fertilisers.

Conclusions

Land and water resources have been identified and characterised as an essential preliminary to agriculture development. The former has been widely used while the latter is still in the stage of initiation in this state. The state has vast potential of water resources. Rivers drain about 183.29 billion cubic metres of water every year. Besides, about 54.9 billion cubic metres are reserved as underground water. Utilising these water potentials, some 120 lakh hectares can be provided with irrigation water (Jain, 1988, p 41), but only 30.89 lakh hectares are irrigated, which is simply one-fourth of the potential. Of course, rivers of the state have been tammed by the bordering states as soon as they leave this state to use their potentials. This is an example of unawareness of the state.

Further, development of irrigation is a localised phenomenon. The disparity in the extent of irrigated area seems to be related with the amount of annual rainfall and its variability to a extent. The quality of land has also been one of the determinants of irrigation, since major irrigated tracts have very good to moderately good capability of land. The socioeconomic conditions have gone long way in creating this disparity. Most often, high intensity of irrigation is in areas of canal and tubewells, and these are in public sector; and as usual, are in politically conscious areas. From this point of view, two classes of people, viz. marginal and small and socially depressed communities have been suffers. Since irrigation has decisive influence on use of other inputs and agricultural productivity, its availability to these people and areas of their dominance is pertinent. In this context, it is suggested that development of minor and even micro irrigation projects would be more economic and useful in the contemporary socioeconomic conditions.

REFERENCES

Balkrishna, S. et al, 1982 : *Socioeconomic Conditions of Drought Prone Areas in Madhya Pradesh*, Hyderabad: N.I.R.D.

Douglas, I., 1975 : Water resources, pp. 57-87 in J.A. Dawson and C. Doornkamp, eds, *Evaluating the Human Environment*, London: Edward Arnold.

Ilbery, B. W., 1978 : Agricultural decision-making - A behavioral perspective, *Progress in Human Geog.*, 2, pp. 448-66.

Jain, C. K., 1988 : *Patterns of Agricultural Development*, New Delhi: Northern Book Centre.

Robertson, V. C., 1970 : Land and Water resource planning in developing countries, *Outlook on Agriculture*, Vol. 6 No. 4, pp. 148-57.

Sharma, S. K., 1987 : Dilemma of resource development and environmental crises in M.P., pp. 236-54 in H.S. Sharma and M.L. Sharma, eds. *Environmental Design and Development*, Jodhpur: Scientific Publishers.

———, 1988a : Regional resource base and pattern of economic development in M.P., India, pp. 123-43 in V. Vidyanath and Mohan Rao eds. *Development of India's Resource Base*, New Delhi: Gyan Pub. Co.

———, 1988b : Adoption of agricultural development measures in relation to size and tenancy of operational holdings in Madhya Pradesh, *Ind. Jour or Reg. Sc.* XX.2, pp. 27-39.

———, 1989 : Use of yield-raising technology in tribal blocks of Madhya Pradesh, pp. 71-80 in P.D. Tiwari and A.N. Sharma, eds. *Tribal Ecosystem and Malnutrition in India*, New Delhi: Northern Book Centre.

Sharma, S. K. and Jain, C. K., 1985 : Changes in agricultural productivity in M.P.: Spatial and temporal dimensions, *The Geographer* XXXII-2, pp. 34-55.

———, 1987 : Green revolution and the faltering steps of agriculture in the Chhattisgarh region, M.P. *Ind. Journ. Landscape System and Ecological Studies*, 10, pp. 41-50.

18

PROSPECTS OF IRRIGATED FARMING IN RAJASTHAN DESERT

BALAK RAM AND A. K. SEN

WATER IS the main problem in the desert. Its proper utilisation and management is essential for perennial farming. Rajasthan constitutes about 62 per cent of the total arid zone of India. Irrigated farming in this fragile ecosystem is still of little significance. However, recently a significant expansion of irrigated farming has taken place through exploitation of ground water and construction of reservoirs and canals. Problems of salinity/sodicity, lowering of water table, seepage and water logging become a matter of great concern not only for farmers but scientists, planners and decision-maker. To overcome this situation an appraisal of present irrigated farming system, its changes over time and space and scope for future development has been made. Resultant problem areas have been identified, delineated and mapped and their impact on crop production and cropping pattern has been assessed. Potential areas for further development are pinpointed and ways and means for judicious use and management of this valuable liquid are advocated. The study is primarily based on the recently developed remote sensing techniques and interpretation of survey of India toposheets, supported by field survey and other available information.

Environment

The Indian desert forms a part of the 'Thar desert' which extends over Pakistan also. In India this desert is spread over parts of Haryana, Punjab, Rajasthan and Gujarat. In Rajasthan the desert is prevailed in its 62 per cent of the total arid zone area of western part (196150 sq.km. area) covering 11 districts viz. Barmer, Bikaner, Churu,

Ganganagar, Jaisalmer, Jodhpur, Jalore, Jhunjhunu, Nagaur, Pali and Sikar. The region is characterised by low and erratic rainfall, extreme variations of diurnal and annual temperatures and high evaporation. The average annual rainfall varies from 456 mm in northeast to less than 100 mm in western most part of Jaisalmer district. The coefficient of variability of rainfall varies from 40 per cent in Jhunjhunu region to 70 per cent in the west (Ramakrishna, 1968). Mean aridity index is 78 per cent (Krishnan, 1964). Probability of occurrence of drought varies from 50-60 per cent (Ramakrishna, 1968). Mean moisture index varies from — 59.5 in Sikar to — 88.9 in Jaisalmer. Temperature during long hot summer days goes as high as 50°C while in cold winter mercury falls below –6°C. The mean maximum expected wind velocity is about 30-40 km/ph but can reach as high as 100 km/ph during severe dust storms. Mean relative humidity during July and August ranges between 75-80 per cent, but during winter it varies from 46-56 per cent. South-west monsoon contributes 91 per cent of total annual rainfall (CAZRI, 1979).

Dominant land forms are the sandy plains with varying degree of hummockiness and the dunes (Ghose et al., 1968). These are encountered with few hills and outcrops (of mainly granite, rhyolite, sandstone and limestone) saline depressions and buried channels. Light textured sandy soils cover major part of this region. Desert soils are mostly sterile with a poor fertility status, poor water holding capacity, very low in organic matter, weaker in exchangeable bases, high in pH value and low in nitrogen and phosphate. The soils have high infiltration rate with unsaturated hydraulic conductivity (Dhir, 1977). The vegetation of the region is quite sparse with limited number of xerophytic plants and thorny bushes. Regarding physiography. Sen (1962) observed that the rivers are misfit and soon dies out due to alluvial suffocation. The drainage, except the Luni, is mostly internal. There is no network of integrated drainage system and no controlling level to which the landscape is lowered.

Present Land Use

Irrigated farming is one of the important constituent of land use ecosystem. In Rajasthan desert dry farming which constitute 91.2 per cent of total cropped area and 86.2 per cent of total cultivated land, is a predominant subsistence agricultural systems. Net area irrigated constitute 6.49 per cent of total geographical area; 10.2 per cent of cultivated area and 12.99 per cent of net area sown. The gross irrigated area constitute 8.78 per cent of total geographical area and 16.3 per cent to total cropped area. Runoff farming system constitute a negligible area (about 0.4%). While turning pages of history of agriculture in Indian desert, the art of agriculture came here from middle east after the beginning of neolithic period. The first agriculture was started by Fisher Folk in north of Bay of Bengal whence ideas of agriculture were passed towards western India, wooded steppe and drier regions (Sauer, 1952; Bhardwaj, 1965 and Randhawa, 1980). During Chaleolithic period (3000 — 1600 B.C.) technique of cultivation and irrigated farming was developed. Two types of agriculture appear to have existed in arid zone. In the riverine areas, which were inundated or irrigated, sedentary agriculturist with comparatively large settlements dominated. Near the isolated water points where only small hamlets were developed, cultivation remained confined to small patches around a few wells in favourable plots. Deserted Kalibangan and other sites on the Saraswati valley shows the evidence of ploughed fields, wheat, gram, barley and cotton (Lal and Thapar, 1977). By the end of 10 century A.D. tank and well irrigation were developed. In later Mughal period (1583 AD) 'nadi', wells and tanks were constructed. *Charas* (Persian wheel), *kosita* and *chanch* were main water lifting devices. After independence and formation of Rajasthan State the region was brought to the main stream of national development. Jagirdari and landlord systems were abolished. Dams and canal came into existence. Five year plans and green revolution brought considerable land use change. Still land use development in Rajasthan desert has always been defeated by climatic barriers, meagre and poor natural resources and political instability. It was the growth of population, their needs and struggle against nature which geared up the momentum to landuse cycle (Balak Ram, 1988). Since 1968 the Basic Resources Studies Division of Jodhpur based Central Arid Zone Research Institute is conducting systematic landuse survey and mapping as an integral part of integrated survey in arid and semi-arid regions of India. Since then landuse survey and mapping on reconnaissance, semi detail and detailed levels on recently developed remote sensing techniques and supported with field survey, secondary information and interpretation of topographical maps and aerial photographs is being conducted. The survey and mapping is based on methodology of land use classification system developed for arid zone by Sen (1978). Sen (1974) has first prepared a landuse map of West

Rajasthan based on interpretation of satellite imagery. Based on the study, out of total geographical area cultivated lands (including short and long fallow) constitute 67 per cent. Of this, 45 per cent area is regularly cultivated. Irrigated double cropped area constitute only 3 per cent Marginal lands (13%) with intensity of cultivation below 30 per cent are almost misused as these are not fit for cultivation. Lands with intensity of cultivation 60-80 per cent constitutes 14.5 per cent and lands with intensity of cultivation 30-60 constitute 14.9 per cent respectively. About 25 per cent area is mapped under different wasteland types. These are also used as grazing ground. This includes sandy waste (12%), saline waste (1.2%), stony and rocky waste (1.9%) and rocky and gravelly waste (9.4%). Permanent pastures and forests constitute 4.2 per cent and 1.3 per cent respectively. These are almost degraded lands due to heavy biotic pressure. Settlement and water bodies together constitute 2.9 and mining 0.1 per cent respectively.

Irrigated Farming

Present Situation

About 13550 sq.km. land of Rajasthan desert has now been put under irrigated farming. This constitutes 6.5 per cent of the total geographical area and 10.3 per cent of the total cultivated lands. The gross irrigated area is about 18300 sq.km. Of this 10900 sq.km. or 59.7 per cent is irrigated by canals which are mainly lying in Ganganagar and Bikaner districts. Wells are second largest source of irrigation and constitute 37.3 per cent share of the total irrigated area. There are about 2 lakhs irrigation wells. Tanks contribute only 2.6 per cent of the gross irrigated area. Ganganagar tops in surface irrigation through Ghagghar Canal, Gang Canal and Indira Gandhi Canal systems, which together irrigate about 10 lakh hectare land. Thus about 34.4 per cent geographical area and 50 per cent of the total cropped area of Ganganagar has been covered under irrigated farming. Second largest irrigated area is found in Pali district where about 2 lakh hectare land is under irrigation. Out of the total gross irrigated are Ganganagar constitute 55.6 per cent, Pali 11.5 per cent, Sikar 5.97 per cent, Jalore 9.3 per cent, Nagaur 4.8 per cent, Jodhpur 4.3 per cent, Jhunjhunu 3.6 per cent, Bikaner 3.1 per cent and Barmer 1.8 per cent. Irrigated area in Churu and Jaisalmer is negligible. Through survey and interpretation of satellite data irrigated area of western Rajasthan have been delineated and mapped (Fig. 18.1). It is evident from the study that groundwater irrigation is mainly concentrated along younger alluvial plains of rivers, over buried channels; valley fill regions, buried pediments and interdunal areas. Thus in Pali, Barmer and Jodhpur districts irrigated area is concentrated along younger alluvial plains of Luni and its tributaries viz. Jojri, Guhiya, Bandi, Sukri, Mithri and Jawai. Other areas are Siwana intermontane basin; Mathania and Borunda region of Jodhpur; Merta, Harsor, Kuchaman and Makrana area of Nagaur district. In Bikaner district canal irrigated zone is mainly concentrated in northwestern part between Kharwara, Chattergarh, Pugal, Manaksar and Bikampur. In Ganganagar district most of the region has been brought under canal area excepting south eastern part of Nohar, Baramsar and Pallu region. In Churu district irrigated area is mainly concentrated in southern part of Rajgarh tehsil between Chandgothi, Nesal, Sankhu, Lilavati and Budhawas. In Sikar and Jhunjhunu irrigated area occurs throughout the region except in northwestern parts in Fatehpur, Ramgarh, Bissau, Malsisar and Bishangarh region. Thus it is evident that ground water irrigation is concentrated in the region of more than 200 mm isohyet.

Tanks are also a source of irrigation, but this is not a definite system as they depend on rainfall. Here almost every alternate year is drought year, though tanks or *nadis* are found all over the region as they meet the requirement of water for livestock and human consumption. But tanks used for irrigation are not so much. Most of them are concentrated in Pali and Jalore districts, apart from few in Jodhpur, Nagaur and Jhunjhunu districts. Important tanks are at Jawai, Hemawas, Chopra, Rajsagar, Sardar Samand, Raipur and Kharda of Pali district and Jaswant Sagar in Jodhpur district. Thus Luni basin has 171 tanks with an utilisable yield of 18962 mcft water (Anonymous, 1985). Beds of irrigation tanks are also used for growing rabi crops (Wheat, mustard and barley and also *zaid* crops viz musk melon, water melon and cucumber etc). This happens when there is no sufficient water for irrigation through canals. The extent of cultivation depends on the moisture available. Irrigated area by source is given in (Table 18.1). Indira Gandhi Nahar Project (IGNP) has 204 km long feeder canal. This water is expected to irrigate about 1.55 million hectares in two stages. During 1986 an area of 2.2 lakh hectare during kharif and 3 lakh hectare during rabi was irrigated by IGNP. That is about 79 per cent of total irrigation potential of stage I available at present.

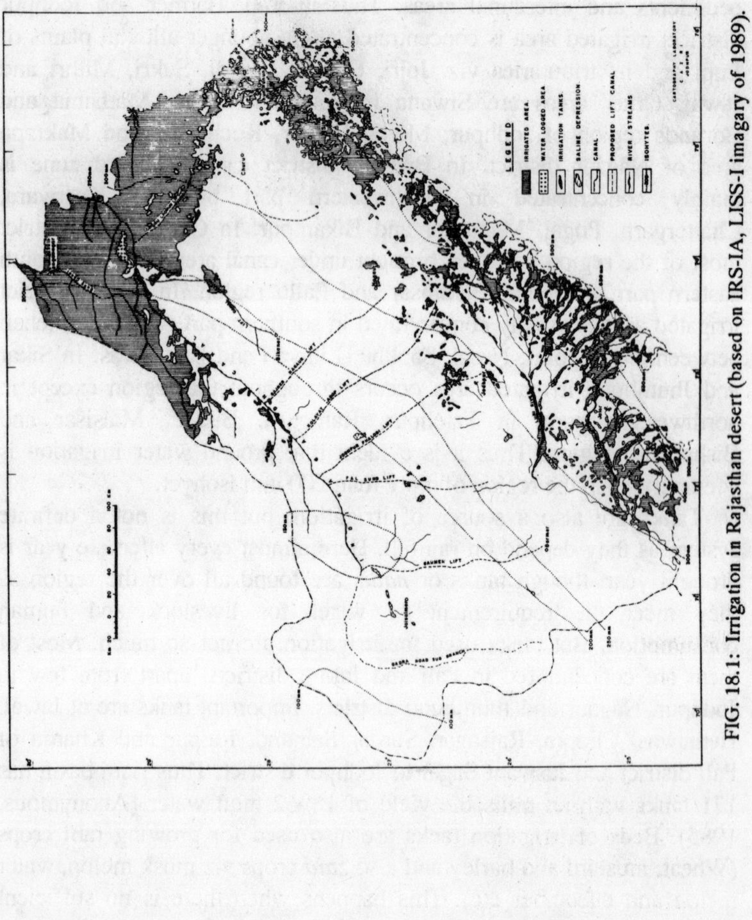

FIG. 18.1: Irrigation in Rajasthan desert (based on IRS-IA, LISS-I imagary of 1989).

TABLE 18.1: Irrigated area (in hectares) by source in Rajasthan desert (1983-84)

District		Wells	Tanks	Canals	Others	Total
Bikaner	N	21	—	37001	—	37022
	G	21	—	56008	—	56029
Barmer	N	18606	—	75	375	19056
	G	32094	—	75	1113	33282
Churu	N	1271	—	—	—	1271
	G	1382	—	—	—	1382
Ganganagar	N	882	—	707891	2	708775
	G	1434	—	1017725	2	1019161
Jaisalmer	N	37	—	1	—	38
	G	37	—	3	—	40
Jalore	N	133705	43	9195	208	143151
	G	160603	43	9957	208	170011
Jhunjhunu	N	55712	32	—	—	55744
	G	65382	256	—	—	65638
Jodhpur	N	52638	23	6015	—	58676
	G	71815	23	6459	119	78416
Nagaur	N	65998	333	342	336	67009
	G	86840	333	426	948	88547
Pali	N	125803	41016	1599	3828	172246
	G	153142	46219	1785	5314	206460
Sikar	N	88156	—	263	—	88419
	G	108994	—	313	—	109307
	Net	542829	41447	762382	4749	1351407
Total	%	40.2	3.1	56.4	0.3	
	Gross	681744	46874	1092751	7704	1829073
1958-59		133923	95796	331207	1031	561957
Changes %		(+409.1)	(-51.1)	(+229.9)	(647.2)	(+225.5)

N = Net irrigated area
G = Gross irrigated area

Cropping Pattern

Choices of irrigated crop depends on the quality and quantity of water and soil characteristics. From canals in Ganganagar district both rabi and kharif crops are taken, whereas from wells and tanks mainly rabi crops are taken. Here application of irrigation to kharif crop is done only to save them from drought stress conditions. Out of total gross irrigated area wheat is the dominant irrigated crops constituting 32.6 per cent; and this is grown all over the region. This is followed by mustard (19.5%) and cotton (17.9%). Cotton is mainly produced in Ganganagar district which account 30.3 per cent of the gross irrigated area of the districts. Rice, groundnut and sugarcane are exclusively cultivated in Ganganagar and Bikaner district out of canal water. Other crops are gram (7.4%), barley (2.4%), bajra (2.4%), cumin (2.1%) and isabgol (1.6%). Cumin and isabgol (2.1%) are exclusively produced in Jalore, Barmer and Nagaur districts. These crops require minimum water for irrigation and can be taken even on undulating lands with sandy soils. Other crops are taramira, chillies, onion, garlic, methi, maize, moong, jowar, alsi, castor, tobacco, potatoes, coriander, sunhemp and fodder and vegetable crops (Table 18.2). Methi has now become more promising crop in Jhunjhunu and Sikar districts. Chillies are dominant in Jodhpur district.

Changes in Irrigated Area

The last three decades have witnessed a significant change in irrigated farming. Since 1958-59 the net and gross irrigated area have increased by 140 and 225.5 per cent respectively. Extent of irrigation has significantly increased in Ganganagar and Bikaner districts due to IGNP. In Ganganagar this has increased from 3.8 lakhs to 10.2 lakh hectares. In Bikaner canal water for irrigation was made available after 70-71. Before that irrigated area by wells was limited to few acres only. Now 37000 hectare land has been put under irrigation with gross irrigated area of 56000 hectare. No specific change has occurred in Churu and Jaisalmer districts because of their geo-physio-climatic constraints. Jhunjhunu, Sikar, Nagaur and Jalore districts gained significant change to the tune of 464, 322, 266 and 276 per cent respectively in net irrigated area, due to construction of new wells. In Pali and Barmer districts significant change has occurred in gross irrigated area. This has happened due to deepening of existing wells and construction of more wells in existing irrigated lands. This is also due to saline ground water conditions and over-exploitation which led to lowering of water table. In Jodhpur district most of the irrigation potential zone was already put into operation. Here net and gross irrigated area have increased by 154.6 and 207.7 per cent respectively. Districtwise change in net and gross irrigated area has been shown through diagram (Fig. 18.2).

TABLE 18.2: Irrigated area under important crops (in %) to the gross irrigated area in Rajasthan desert (1988)

District	Bajra	Wheat	Barley	Gram	Rice	Mustard	Taramira	Ground-nuts	Cotton	Sugar-cane	Chillies	Onion	Jeera	Isabgol	Maize	Other
Barmer	2.3	33.0	0.7	0.2	—	25.7	0.2	—	—	—	0.1	0.4	31.4	31.4	0.1	4.9
Bikaner	0.6	20.2	—	17.5	—	19.1	1.4	8.2	1.0	0.5	—	—	—	—	—	31.5
Churu	7.8	47.6	2.5	8.7	—	32.2	0.7	—	—	—	—	—	—	—	—	0.5
Ganganagar	0.8	30.2	0.5	9.5	2.1	10.4	0.2	0.2	30.3	0.7	0.1	—	—	—	0.1	14.8
Jaisalmer	—	33.0	2.0	—	—	15.0	—	—	—	—	—	—	—	—	—	50.8
Jalor	1.4	23.7	0.9	0.4	—	25.5	0.1	—	0.7	—	0.5	0.2	16.2	16.8	0.1	13.5
Jhunjhunu	7.3	44.7	4.1	13.7	—	18.4	0.6	—	0.1	—	0.9	1.3	—	—	—	8.9
Jodhpur	1.6	37.4	0.5	—	—	29.0	0.3	—	1.9	—	10.4	1.0	—	—	0.1	8.9
Nagaur	10.2	31.4	6.3	0.8	—	33.1	0.1	0.1	3.4	—	0.9	0.7	0.2	0.2	—	12.4
Pali	0.2	48.3	7.0	1.4	—	26.4	0.3	0.1	6.0	—	0.7	0.1			2.9	6.6
Sikar	14.9	35.9	10.9	13.2	—	11.0	0.1	—	0.1	—	0.8	0.8	—	0.1	0.1	12.2
Total	2.4	32.6	2.4	7.4	1.2	19.5	0.4	0.3	17.9	0.4	0.8	0.2	2.1	1.6	0.4	10.4

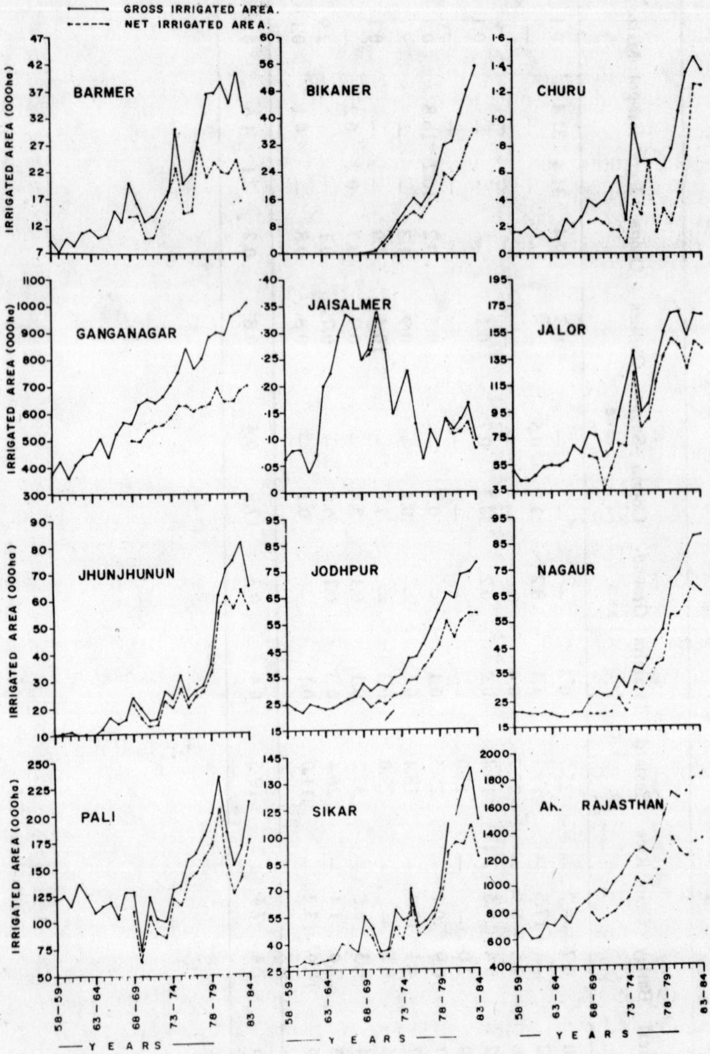

FIG. 18.2: Trends in net and gross irrigated area in Rajasthan desert.

Changes in Source of Irrigation

During the last 25 years the area irrigated by wells has increased by 409.1 per cent, canals 225.5 per cent and other sources 647.2 per cent, although it has declined to the tune of 51 per cent in case of tank irrigate. This is due to frequent droughts in recent years and also because the tanks do not get sufficient water for irrigation even though new irrigation schemes have come into existence. Thus irrigation from tanks is never certain and no planning can be based on such sources. The old water lifting devices like Persian wheel and charas have almost vanished and most of the wells have diesel engine or electric pumpset. Persian wheels still exist along the Aravalli foot hill zone where water table is high.

Changes in Cropping Pattern and Crop Production

Increased irrigation farming has certainly changed the agriculture economy by way of getting good production from improved seeds and introduction of cash crops. Thus most of irrigated crops are cash crops and for other cereals or pulses irrigation is given only to save the crop from stress period and droughts. Based on data of irrigated crops and their production for the year 1956-59 and 1983-84 the changes have been worked out. The area of rice, wheat, mustard, groundnut, cotton and sugarcane has increased by 586.7, 58.6, 452.5, 679.0, 379.2 and 64.6 per cent respectively. No specific change has occurred in gram, while area and production of barley has declined by 71.6 and 67.8 per cent respectively. As regards production, it has shown significant change due to improved agricultural practices and high yielding varieties. During this period the production of rice, maize, wheat, mustard, cotton and sugarcane has gone up by 2424; 100.8, 148.4, 863.9, 1050.8 and 343.7 per cent respectively. The production of chillies has also increased by 336.6 per cent whereas its area has increased by 23.6 per cent only (Table 18.3). While looking for trend in cropping system under irrigated condition it has been observed that cumin and isabgol have become more promising crop in Barmer and Jalore; wheat and 'rizka' (Fodder crop) in Bilara and chillies in Mathania region of Jodhpur district; maize and mustard in Pali; and methi in Jhunjhunu and Sikar districts. Groundnut, cotton and sugarcane have mainly come up in canal irrigated area of Ganganagar district. Thus, in ground water irrigated areas salt tolerant crops and crops which require less water are being given more impetus. The districtwise area and production figures of principal crops (both under irrigated and non-irrigated condition) during 1987-88 reveal that under rainfed conditions though area has increased but production has not shown any significant result (Table 18.4). The reason behind this is that the growing population has brought more marginal lands under cultivation. It is also clear that high yielding varieties of bajra, moong and sesamum require more moisture and are not susceptible to droughts and these too need irrigation. Wheat, mustard, taramira and gram are also produced without irrigation — through conserved moisture and runoff farming system. In the socalled 'khadins' water is collected during monsoon period and after it dries up, the rabi crops are produced. In Jaisalmer district rabi crops are mainly produced through this farming system. There are 18 big khadins with irrigation capacity of 667 hectare.

Consequences of Irrigated Farming

Irrigated farming in long run is not always fruitful, particularly in arid regions where geological formation, soils, climate, and groundwater play an important role over agricultural land use. Erratic and highly variable rainfall, extremes of temperature, saline and deep seated ground water, absence of perennial streams and low fertile soils are main constraints of arid zone farming. Water erosion and erosion and deposition of sands by wind are other associated problems.

Salinity/Sodicity

While discussing the productivity of arid lands for farming, Raheja and Mishra (1964) pointed out the main consequences of irrigated farming. Arid zone soils hardly have any structure, and are in low fertility status. The constant cropping on agricultural land without return of nutrient to this soil leaves it exhausted. Thus soil water aeration environment is not conducive to high rate of production. The soils become saline by use of saline water. As a consequence the crop yield is appreciably reduced. The salinity has more impact on irrigated farming rather than on rainfed agriculture. The excessive and unjudicious use of water having moderate to high residual sodium carbonate cause sodicity problems. In Balotra area comprising Budiwara, Meli, Khakharlai, Kusip and Janiyana continuous

TABLE 18.3: Changes in area and production of major irrigated crops in Rajasthan desert during 1956-57 to 1983-84 (Area in hectares, production in MT)

Crop	1956-57		1983-84		Changes	
	Area	Production	Area	Production	Area %	Production
Rice	3076	2360	21122	59568	586.7	2424.4
Maize	24000	18000	37060	35140	54.4	100.8
Wheat	420480	433298	666000	1051400	59.0	145.0
Barley	129291	410591	36740	45280	-71.6	-67.8
Groundnut	2000	NA	15581	12231	679.0	—
Mustard	63839	24338	352700	234600	452.5	863.9
Castor	NA	NA	3170	735	—	—
Cotton	86000	44000	412120	506368	379.2	1050.8
Sugarcane	3976	87205	6543	386901	64.6	343.7
Chillies	6107	3029	7549	13226	23.6	336.6
Gram	610000	479000	886200	427300	—	—
Tobacco	—	—	595	1061	—	—

TABLE 18.4: Area and production of principal crops in West Rajasthan (1987-88)

District	Rice		Jowar		Bajra		Maize		Wheat		Gram		Groundnut		Sesamum		Mustard		Cotton	
	A	P	A	P	A	P	A	P	A	P	A	P	A	P	A	P	A	P	A	P
Barmer	—	—	1214	301	533564	—	135	50	11725	22244	5	3	—	—	3082	15	236	110	22	14
Bikaner	61	51	491	122	170877	1242	25	9	9150	17359	11840	7115	12582	11026	2724	—	9830	7632	4826	3059
Churu	—	—	19	5	316633	14421	1	—	1169	2269	49946	403	1	1	110	3	814	693	—	—
Ganganagar	17693	27473	511	127	38837	282	292	109	198536	376641	148913	89488	44583	6357	35	1	141966	117209	297753	193248
Jaisalmer	—	—	11	3	28567	—	—	—	542	1028	17	10	—	—	34	1	342	119	—	—
Jalor	—	—	944	234	174991	22678	148	74	32131	38297	18	11	1	1	8380	17	1587	851	615	390
Jodhpur	—	—	66	16	249251	32317	2	1	29742	44614	7246	7151	14	7	34	1	36	35	5	3
Jhunjhunu	—	—	12630	361	460327	8961	29	11	18503	25951	30	18	47	23	17642	35	79519	6700	1173	743
Nagaur	—	—	81455	—	468315	17853	369	138	29681	44522	727	437	3572	757	32761	131	7360	5390	1642	1041
Pali	1	1	99139	24592	135717	17597	21587	8047	27372	51927	206	124	662	—	56651	113	15975	13470	5842	3703
Sikar	1	1	134	33	215637	17488	129	48	41276	77702	18593	18099	713	353	309	8	36514	16877	46	29
Total	17756	27526	196614	25974	2792626	132839	22717	8487	399827	702554	237641	122859	62175	18525	121762	325	294210	169086	311924	202230

A = Area in hectares P = Production in Tonnes.

irrigation with water of 6-20 me/1 RSC the soils have developed high sodicity (pH 8.8 to 9.7) (Dhir and Joshi, 1985). The soils have become very hard and compact and infiltration has been greatly reduced (Ram, 1988). Now farmer can grow only rabi crops once in two to three years. Such problems have also developed in Ahor tehsil of Jalore district, Sojat and Pali tehsils in Pali district, Doli and other villages towards south of Jodhpur, Bilara area and southern part of Nagaur district. This problem has further engulfed the flood plains of Mendha river which are being irrigated with saline water. Shankarnarayan (1983) observed that salinity problem (both natural and man induced) are dominant in western part of upper Luni basin. The production and intensity of cropping of lands is gradually declining.

Lowering of Water Table

The entire ground water irrigated zone is subject to the problem of lowering of water table, decline of yield due to consecutive droughts, non-recharge of aquifers and over exploitation of ground water. From all existing wells constructed on 10 years back, the water has gone down from 5 to 20 metres during the last 25 years. That means every year the water table has receded by 2 to 4 feet. Those who can afford the deep boring can get some water, while others do not get water for irrigation. In Kantaliya river alluvial fans in Rajgarh tehsil of Churu district, the irrigated area has not shown any significant increase during the last 25 years. Few new wells have been constructed on the one hand, the yield from a number of wells decreased on the other. Here water table has gone down from 23 to 46 m during 1967-90. This problem is more precarious in valley region of north eastern parts of Sikar and southeastern part of Jhunjhunu districts. As compared from multidate satellite imagery and ground verification, double cropped areas of this region are vanished. Irrigated crops can only be seen in pockets and that too from new wells. It is evident from the fact that about 14000 wells are lying dry in Sikar and Jhunjhunu region.

Seepage and Waterlogging Problems

Through the interpretation of landsat (FCC) imagery the waterlogged areas caused by seepage along canal irrigated region were identified and mapped in Suratgarh. Hanumangarh, Tibi and Anupgarh areas of Ganganagar district (Fig. 18.1). Sen (1986) found a submerged 20 sq.km. area due to waterlogged problem near Suratgarh. Due to irrigation the water table has been considerably increased around Suratgarh, Baropal and Rangmahal. The inundated areas which were waterlogged as early as 1912 have become saline and this particular problem is gradually increasing towards the south in recent years. Thus proper management, distribution and drainage system are essential to overcome such problems which are posing threats in the years to come.

Other Problems

Looking into the extent and nature of irrigation, status of groundwater exploitation and the problems caused by unjudicious and excessive use of groundwater; use of saline water and irrigation with water having high residual sodium carbonate, mismanagement of irrigation water, inadequate and improper watering and cropping system, it has now become essential to examine the problem in depth keeping in view the all biophysical resources; climatic factors and other environmental condition, and suggest proper use and better management practices to get constant enhanced agricultural production and also pinpoint the potential zones for further exploitation and conservation of ground and surface water resources. ORG (1988) study pointed out that at present water distribution in most of the irrigation projects is characterised by high level of operation loss, inequality in volume of water supplied to different zones, supply being not available in right quantity and right time. For this, development of appropriate water allocation strategies is needed.

Irrigated Farming : Potential and Management

Irrigation Potentials

Dhir and Krishnamurthy (1952) have estimated 0.42 m acre ft. as available surface water in Luni basin. Irrigation Department of Government of Rajasthan (1984) has estimated the irrigation potentials of Luni basin, as 100,000 hectare through 165 existing tanks, six tanks under construction and 36 proposed tanks. Total yield available is about 30290 mcft. Total yield utilisable is 23218 mcft and balance of 7072 mcft yield is in basin. Dasgupta (1977) estimated that 1068 mcm of usable water can be exploited annually for meeting the drinking water need of livestocks and irrigation requirements. He further maintained that total mineable groundwater storage from some of the aquifers of Jaisalmer, Barmer and Bikaner are so high that even dewatering of few metres of aquifers can meet the requirement for various purposes for next 200 years. The districtwise exploitable potential in mcm is (Barmer 75.0), (Bikaner 22.3), (Churu 53.9), (Jaisalmer 137.7), (Jalore 168.6), (Jhunjhunu 129.8), (Jodhpur 98.05), (Nagaur 224.9), (Pali 77.9) and (Sikar 96.8 mcm). From the Indira Gandhi canal the water is expected to irrigated about 1.55 million hectares in two stages. The total flow is 2.225 m acre ft during kharif and 1.916 m acre ft during rabi. The proposed lift canals are meant to provide drinking water for human and livestocks raising nurseries and plantations, apart from irrigation. Thus, after completion of stage II of IGNP canal system, it will not only enhance agricultural production but also help in settling the population and development of silvi-pastoral system, grass lands, forage production and communication net work. Through systematic resistivity survey more groundwater potential zones can be delineated and groundwater can be tapped with financial and technical support to the farmers. Geomorphological studies conducted in CAZRI (Ghose et al 1979; Kar, 1986) also reveal the buried channels as most potential sites for groundwater development.

Management Practices

Irrigated farming in arid Rajasthan need proper management, judicious use, identification and proper exploitation of groundwater resources, optimum water harvesting and rational planning for years to come. Some suggestions for different irrigation practices are given here.

A. *Canal Irrigation*

Canal irrigation needs the following:

1. Minimise operation loss
2. Proper supply of water at proper time and in appropriate quantity
3. Adequate drainage
4. Proper maintenance and surveillance
5. Prevention of seepage and waterlogging
6. Evolve suitable cropping pattern and crop rotation and development of horticulture

7. Prevention from getting canal buried with sand deposition.

B. *Groundwater Irrigation*

1. Judicious use of groundwater in view of its quality and soil characteristics
2. Minimise transmission loss from irrigation channels by percolation and evaporation
3. Improved and efficient methods of irrigation viz. drip, sprinkler and trickle systems
4. Growing salt tolerant crops and crops which require less water
5. Proper choice of crops and crop rotation
6. Substantial production can be achieved with less water by optimising the inputs
7. Application of irrigation at the rate of potential evapotranspiration and application of additional water for leaching
8. In case of double cropping, application of phosphorus through winter crops is essential. Application of heavy dose of nitrogen is required since the nitrogen is chronically deficient in the soils of this region
9. Proper fallowing system should be maintained to permit the leaching of salts in case groundwater is saline
10. Application of gypsum and fallowing is needed where water has high residual sodium carbonate
11. Application of saline water mixed with sweet water at the time of germination or fresh water can be applied at the time of germination and later on saline water can be supplemented.

C. *Runoff Farming*

1. Construction of khadins on suitable sites
2. Construction of farm pond to recycle collected water and provide supplement irrigation during water stress
3. Application of surface multches to check excessive evaporation of moistures from soils
4. Deepening of irrigation tanks and save them from siltation and pollution
5. Proper drainage system to avoid salinity problem.

Conclusion

The study revealed that during the last three decades, irrigated area has significantly increased and crop production has also gone up. Irrigated farming has mainly increased due to exploitation of more ground water and Indira Gandhi Nahar Project. Irrigation from tanks has been declined due to frequent droughts and increased human consumption. Changes in cropping pattern has also occurred. More impetus is now being given to cash crops and the crops which require less water. Excessive use of saline and high RSC water at places caused salinity/sodicity problems which has ultimately declined the productivity and fertility. Lowering of water table has been observed in all parts of ground water irrigated lands. In Sikar and Jhunjhunu where irrigation is expanding on sandy soils, the valley regions have witnessed decline in irrigated area and thousands of wells are dried up. Seepage, waterlogging, rising of water table and operation loss are main problems of canal irrigated lands.

There are still a number of groundwater potential zones which can be identified and tapped. Judicious use of water, efficient methods of irrigation, prevention of conveyance loss, proper management and growing of suitable crops and crop rotation together can give optimum production and sustained agricultural economy. Canal irrigation can be appropriately managed through suitable strategies. Thus irrigated farming has vast potentialities, but it needs wise and judicious use and management of both surface and groundwater, supplemented with rational planning, political will and peoples awareness.

REFERENCES

Anonymous, 1979 : *Climate of western Rajasthan*, CAZRI, Div. TECH BUL 1/79.

Anonymous Statistical Abstract of Rajasthan, 1956-57 to 1985, Economic and Statistical Deptt., Govt. of Rajasthan.

Anonymous, 1984 : Master Plan of Lunib basin, Irrigation Department, Govt. of Rajasthan.

Bajpai M. R. and Michael A. M., 1964 : Improving irrigation in the Arid Zone, *Proc. Sym Problems of Indian Arid Zone*, Jodhpur, pp. 208-10.

Bhatnagar, N. K., 1988 : Sankhikiya Roop Rekha 1985 for Barmer, Bikaner, Churu, Ganganagar, Jaisalmer, Jhunjhunu, Jodhpur, Nagaur, Pali, Sikar and Jalor district, Economic and Statistical Director, Govt. of Rajasthan.

Das Gupta, K., 1977 : Ground water resources *Desert Eco-system and its Improvement, CAZRI*, pp. 26-30.

Dhir, R. P., 1977 : 'Soils' Desert Ecosystem and its improvement, CAZRI, pp. 115-18.

Ghose, B. Kar, A. and Hussain Z., 1979 : Lost coarse of the Saraswati river in the Great Indian Desert — New evidence from Landsat imagery, *Geographical Journal*. Vol. 145, pp. 445-51.

Joshi, D. C. and Dhir, R. P., 1989 : Management of loamy sand soils irrigated with high RSC water, *Pro Int. Sym Managing Sandy Soils*, CAZRI, Jodhpur, pp. 474-78.

Kar A., 1986 : Remote Sensing of buried farmer streams in the extremely arid terrain of Jaisalmer for water and salinity, *Proc. Seventy Asian Remote Sensing Conference*, Sec B2/1-132/9.

Krishnan, A., 1977 : *Eco-climatological features Desert Eco-system and its Improvement*, pp. 38-51.

ORG, 1988 : *Study of water management strategies in Indira Gandhi Nahar Project*, Operation Research Group, Ahmedabad.

Lahiri, A. N., 1977 : Physiogeological Aspectsal Crop Yields in Relation to Water Stress Desert Eco-system and its Improvement, CAZRI, pp. 193-201.

Rahija, R. C. and Mishra, D. K., Productivity of Arid lands for farming and Animal husbandry, *Proc. Sym. Problem of India Arid Zone*, Jodhpur, pp. 472-77.

Ram, B., 1988 : *Recent changes in landuse in an arid environment, A case study of Siwana region*, (Unpublished Ph.D. Thesis).

Ramakrishna, Y. S., 1985 : *Climatic Changes in Relation to Desertification in the Indian Desert.*

Saxena, S. K., 1977 : Vegetation of Indian Desert Biomass and Primary Productivity of Grass lands, *Desert Eco-system and its Improvement*, pp. 144-71.

Saxena, S. K., 1985 : *Desertification in western Rajasthan.*

Sen, A. K., 1962 : Does Siwana and its Evolution to an Arid Cycle, *The Indian Geographer*, Vol. 7 No. 1, pp. 78-84.

Sen, A. K., 1985 : *Environmental Degradation due to Desertification in Rajasthan Desert.*

Shankarnarayan, K. A., and Kar, A., 1983 : *Natural and Human Resources Appraisal of Upper Luni Basin.*

Singh, R. P., 1977 : Dryland Farming Technology *Desert Eco-system and its Improvement* CAZRI, pp. 202-12.

Thanvi, K. P. and Krishnan, A., 1975 : Study on the Variability of Rainfall and Evaluation of Major Rainfall Type in W. Rajasthan, *Proc. Workshop on Problem of the Deserts in India*, GSI pub. No. 49, pp. 156-60.

19

FERTILISERS, MANURES AND URGENCY OF NATURAL FARMING

LAKHVINDER SINGH

SOIL LAYER, that is capable of supporting flora and fauna has evolved through a long geomorphic process. The fertility of the soil is usually expressed in pH value, which is assessed by soil nutrients. These nutrients are macro and micro, present in larger and smaller amounts. Macro nutrients like carbon, hydrogen and oxygen are derived from air and water. Micro nutrients like nitrogen, phosphorus, potassium, calcium, magnesium, iron, copper, zinc, manganese, baron, chloride and molybdenum are largely obtained from soil.

Quality soils have sound base for better farming systems. Ever since the dawn of civilisation soils have been exploited to raise food crops. With rapidly increasing population, farming practices have undergone a rapid diversification. More pressure is being put on land resources to obtain a rich harvest and increase foodgrain production. This trend has resulted in exhausting soil nutrients considerably. Chemical fertilisers are being applied on a larger scale to obtain better yields and remove deficiency of soils.

Fertiliser Use and Modern Agriculture

Chemical fertilisers find a favourable place in agricultural practices. Countries, world over are using synthetic fertilisers to enhance crop yield. In India too, fertilisers found a commanding place, especially after the second green revolution. Production on the farm front registered increased yields and we were successful in putting a stop to foodgrain imports from western countries.

Fertiliser use is maximum in Uttar Pradesh, Punjab, Andhra

Pradesh, Karnataka and Haryana. Fertiliser consumption was highest in Punjab during 1988 which was 155 kgs per hectare. Yield of various food grains has also increased, but it is incomparable to those of Japan, United States or China (Table 19.1). There is no denying the fact that overall position has improved, that we have quite a surplus for emergency needs. But in Indian conditions we have not been able to get satisfactory results with chemical fertilisers as inputs.

TABLE 19.1: Rice yield for some countries and Indian states — 1986

S.No.	Country	Kg. per ha	S.No.	State	Kg. per ha
1.	Japan	6,225	1.	Punjab	3,221
2.	U.S.A.	6,095	2.	Haryana	2,800
3.	China	5,346	3.	Tamil Nadu	2,490
4.	U.S.S.R.	3,898	4.	Andhra Pradesh	2,210
5.	India	2,179	5.	Karnataka	1,800

Source: Ministry of Agriculture, Govt. of India.

Recent studies, the world over, by eminent scientist and environmentalists has focused on the need to preserve land from the ill effects of fertiliser use. Recent finding bring out that fertiliser use is restricted only to those areas where assured irrigation facilities exist. Irrigation is mainly by canals and tubewells, both electric and diesel operated. Irrigation extension works are largely concentrated in the fertile plains of major agriculturally developed regions over the globe. Better results have been realised in certain regions, due largely to assured supply of irrigation and in regions where virgin soils have been exploited by its use.

It is estimated that more than 55 per cent of nitrogenous fertilisers so added to the soil, to increase crop yields are lost. Only a small portion is utilised by straw and grain for its growth (Table 19.2). A certain proportion of the fertiliser added to soil is either washed or drained away. Major part of which finds its way to ground water, rivers or lakes, contaminating potable water supplies. Thus promoting health hazard. Nitrogenous fertilisers only help in temporary growth. Their effect on plant growth is spontaneous and quick. Thus it works as a booster for crop culture. Their use over longer periods makes plants immune to fertiliser use. Thus curtailing agricultural output.

Continued use of chemical fertilisers makes soil barren and desertic. Soil characteristic undergo a great change. Humus in soil is lost. As a consequence Micro-organisms are adversely affected, with the result, that, micro-nutrients in soil do not get replenished to enhance soil fertility. Their ill effects on human health have also been established. Water contaminated by nitrates enter human body through food chain where it is converted to nitrites which bind hemoglobin thus impairing oxygen carrying capacity in blood which may prove fatal. Harmful effects have been detected, relating to various human diseases. They are much more dangerous for children and infants.

TABLE 19.2: **Recommended dose of fertiliser's and nutrients removed by some field crops**

Crop	Recommended Dose Kg/ha			Nutrients Removed* Kg/ha		
	N	P_2O5	K_2O	N	P_2O5	K_2O
Paddy	100	50	50	83	24	125
Maize	100	75	75	116	48	105
Cotton (irrigated)	10	50	50	32	17	45
Groundnut	25	50	25	80	23	45
Wheat (irrigated)	110	50	50	60	30	68
Linseed	25	25	—	19	12	33
Sugarcane	250	115	115	118	72	357

* by straw and grain

Source: Compiled from various sources.

The data for the consumption of chemical fertilisers is very much on records. But no feasible data exists as to the negative impact of fertilisers on soil fertility, soil characteristics, water contamination, ecological hazards, and environmental degradation. We have tried to bring out a correlation between chemical fertiliser use and increase in yields. India has been able to double its food grains production since 1960-61 to 1988-89. But at the same time more areas have been brought under the plough. This had been possible not by the use of chemical fertilisers, but also as a consequence of extension in irrigation, more inputs in the form of power (electric as well as diesel), high yielding variety seeds, mechanisation, technical advancements and large agricultural labour force.

The agricultural heartland of India comprising Punjab, Haryana and Western Uttar Pradesh is a testimony to this effect. But at the same time we also admit negative impact of fertiliser use. A very lively example is the dust bowl of United States of America, which has been transformed from lush green fields to desertic land.

Impact of Fertilisers, Allied Infrastructure and Agricultural Dimensions: Agricultural Heartland Scenario

Increased outputs in food grain production exceeding generalised targets have been achieved in India's 'Agricultural Heartland'. The heartland comprising western Uttar Pradesh-Haryana-Punjab belt. This could be made possible by large infrastructure base existing in these states. The region is economically better, farmers are adoptive to new technologies in agricultural field, they are more enterprising than their counterparts in rest of the country, coupled with network of irrigational facilities, fertilisers, high yielding variety of seeds, pesticides, intensive use of agricultural machinery, farm management practices and fertile soils of Sutlej-Ganga Plain.

The region owns largest agricultural machinery which is more than 45 per cent of India's total agricultural machinery. Punjab owns highest number of working tractors and harvestors. It owns, 1,44,466 tractors (27.97%) as compared to Uttar Pradesh's 1,09,642 (21.23%), Haryana's 36,229 (7.03%), compared to all India's total of 5,16,461 tractors respectively (Table 19.6).

Irrigation practices are more advanced, the network of canals is more than a hundred years old, unique and largest of its kind in the world's agricultural economy. With recent advancements in science and technology, pump sets, both electric and diesel operated have fast multiplied during the last decade. More inputs of diesel and lubricants are being applied to raise food and commercial crops. Punjab, Uttar Pradesh and Haryana own 4,52,000, 6,52,000 and 1,93,000 pumpsets, both electric and diesel operated respectively as against all India's total figure of 47,97,000 pump sets.

Chemical fertilisers found an easy access to this region. Use of fertilisers has registered an increased application during the second green revolution. Fertiliser consumption is highest as compared to all india average of 53 kg per hectare. Large inputs of fertilisers are being siphoned into soil for increasing farm output. Punjab applied higher doses of fertiliser inputs, per unit area. Consumption during 1988 in Punjab was 1112 thousand tonnes as compared to 1500 thousand tonnes in Uttar Pradesh and 394 thousand tonnes in Haryana.

All the above cited factors helped turned this region, into grannary of India a much prestigeous and coveted position amongst all states of the Union of India. Agricultural production has registered an all time high growth rate and more specially in Punjab; wheat 3580 kg, rice 3160 kg, barley 2289 kg and maize 1000 kg per hectare, compared to India's yield of, wheat 2998 kg, rice 1482 kg, sugarcane 5900 kg, barely 1387 kg and maize 378 kg per hectare. Thus establishing a coveted position in the Indian Republic on the basis of per unit area production.

With additional doses of inputs the region has achieved higher production trends. But now a near saturation level is being perceived with respect to foodgrain production in agricultural belt. Additional doses of inputs are not giving desired results. The phenomena of diminishing returns is being realised. Additional capital investments are not generating satisfactory results. Surpluses on farm front are rapidly narrowing down.

Addition of agricultural inputs have adversely affected the soil fertility. No time is allowed to replenish nutrients in a natural way. Organic and farm yard manures are no longer applied to enhance soil nutrients. Water table is going down and water supplies are being contaminated by the use of chemical fertilisers and pesticides. Soil scientists estimate that two-third of Punjab, Haryana and Uttar Pradesh may turn barren and desertic land, if not properly attended to.

It is binding and obligatory on man to use natural resources i.e. soils diligently and prudently, allowing time for regeneration. Population is fast increasing and so are the demands for food, fuel fodder and fibre and additional land to build upon. There need be enough interaction between climate, geosphere and soil. Soil is being modified by technological and scientific innovations, thus changing landuse pattern. Soil can be made rich by micro-nutrient culture, bio-fertilisers, biological control of pests and step towards natural farming practices to conserve soil.

Do fertiliser, inputs induce better output?

We all are aware of epoch making impact of green revolution and consequent foodgrain output. But do the chemical fertilisers and pesticides really help in increasing foodgrain output is a vexed question. The paper published by the Punjab-Haryana Development (PHD) chamber of commerce and industry on 'Agriculture production trends', has observed "that the average agriculture growth rate has declined in the post-green revolution period. Post green revolution period shows a growth rate of 2.8 per cent per annum, whereas pre-green revolution period 1949-50 to 1964-65 showed an annual growth rate of 3.0 per cent. A period when chemical fertilisers were unknown to Indian farmers. Peak to peak growth rate during the 80's is, however, likely to be 2.8 per cent (assuming 1 per cent increase in agricultural production in 1989-90 over 1988-89), which is still lower than the average annual growth achieved in the pre-green revolution period.

Organic Manures : Their Characteristics and Adaptability

Now that we are self sufficient in food grain production, we need to improve the characteristics and nutrients of our soils, which will also improve quality of our environment. Taking gross account of the harmful effect of chemical fertilisers, it is urgent and desirous to slowly and steadily switch over to organic manures and make a shift towards natural farming practices. Various organic manures are:

(i) Farm Yard Manure (FYM)
(ii) Green Manures
(iii) Concentrated Organic Manures
(iv) Bio-fertilisers and
(v) Poultry waste

Manures are rich in soil nutrients and provide food to millions of soil micro-organisms, which in themselves are good and ready source of soil nutrients after their death and decay. Thus in a way contributing to soil fertility and also regenerate Physio-chemical properties of the soil.

Manures are generally decomposed before being applied to the fields. They contain dung, urine, straw, farm waste, household waste and various other organic and inorganic materials (Table 19.3). They are prepared with great care, soil nutrients are preserved from going waste of escaping to air or water.

Farm Yard Manure

Ten Tonnes of FYM/acre supplies 45 kg nitrogen, 19 kg Phosphate and 45 kg of Potassium, F.Y.M. contains 0.5-1.5 per cent N, 0.4-0.8 per cent P_2O and 0.5-0.9 per cent K_2O. It also contains various micro-nutrients essential for maintenance of soil fertility. Unlike chemical fertilisers FYM releases soil nutrients for plant growth in a slow but continuous process. Thus the nutrients for plant growth are available for full period of growing season, and some nutrients are retained for next crop also.

India's large cattle wealth contributes large quantity of dung and urine. Annual production per year is nearly 7300 kg per animal. Similarly human excreta i.e. night soil is having more NPK value than the animal excreta and fodder waste. In towns and cities a huge quantity of sewage and waste products are good source for manure when decomposed.

Manures even supplement the soil nutrients for 3-4 successive years at a stretch.

Green Manuring

Usually thick growing leguminous crops like cowpeas, sun-hemp, bersem, varieties of pulses and Lobia are good source of organic matter and nitrogen. These crops are grown are ploughed in the field at a time when their stem is still green and have attained sufficient height and reasonable green leafy matter. This supplies 8 tons of organic matter, 40 kg of nitrogen and approximately 2/3 of nitrogen is fixed by bacteria living in the root zone of these green manuring crops (Table 19.4).

Concentrated Organic Manures

These includes non-edible oil cake, edible oil cake and other materials like blood meal, fish meal, bones, hoof and horn meal. These are the waste products of mills, factories and butcher houses. They have high nutrient content i.e. cotton seed cake has 6.4 per cent N, 2.9 per cent P_2O5 and 2.2 per cent K_2O.

Bio Fertilisers

Various micro-organisms fix NPK to the soil in a free state. It is a well known fact that soil organisms play a vital role in the maintenance and sustenance of soil fertility, through breaking and decomposing organic matter.

TABLE 19.3: Carbon-nitrogen ratio of various organic materials

Material	C/N Ratio	Material	C/N Ratio
Farm Waste/Residue		*Human/Household Waste*	
Rice Straw	80-130	Night soil	6-10
Wheat Straw	80-130	Digested slug	6
Barley Straw	100-120	Bio Gas Slurry (ex-Dung)	20
Maize Stalk/leaves	50-60	Potato Tops	27
Cotton Stalk	70	Onion	15
Sugarcane Trash	110-120	Carrot	27
Green weeds	13	Turnip	17
Fallen leaves	44	Fruit wastes	35
Grass clipping	40-80		
Animal Waste		*Forest Waste*	
Cow Dung	19	Leaves	40-80
Buffalo Dung	19	Raw saw dust	208
Sheep Dung	29	Rotted Saw Dust	128
Horse Dung	24		
Pig Dung	13		

Source: Compiled from various sources.

TABLE 19.4: Degree of N fixed by certain legume crops

Legume	Nitrogen Fixed (Kg/ha/Yr)
Alfa Alfa	194
Cow Pea	90
Peas	72
Sunhemp	58
Beans	40
Berseem	120
Ground nut	42

Source: Punjab Agricultural University, Ludhiana.

Azotobacter utilise atmospheric nitrogen for their protein synthesis. On their death it is assimilated to soil and contributes toward enrichment of the soil to the extent of 11-14 kg N/ha. Blue and green algae in submerged paddy field can fix 35-65 kg of nitrogen per hectare per year. Similarly Azolla a floating firn, produced biomass to the extent of 35-40 Tons per hectare per year which contains 4-5 per cent nitrogen on dry weight basis. This can help save 30 kg. N per hectare per year.

Poultry Manures

Poultry manures contain highest nutrient level among all organic manures, on dry weight basis is contains 1.46 per cent *N*, 1.17 per cent *P* and 0.62 per cent *K*. Poultry rearing converts non-edible industrial and agricultural products into rich manures (Table 19.5).

Natural Farming for Sustained Agriculture Production

Natural farming system has attracted the vigilance of ecologists, environmentalists and scientists. In Japan, United States and many American and European countries this wave swept the nations in mid-seventies. The more advanced nations has seen the practical importance of natural way of farming. United States and Japan, can now boast of 20-35 per cent more production by using organic manures with great stress on natural way of farming practice.

The essence of natural farming principle emphasises that, we have the right to carry away the grain i.e. the physical out put or produce. We put in seed (grain) to produce this grain. Man has no right over the straw which is a by product. Soil nutrients have been consumed to produce grain, therefore, they must be replenished. The best way to restore is that, straw be allowed to exhume in the field which will supply the required nutrients to the soil. Humus in soil will increase which will help multiply micro-organisms which in due course would help restore soil fertility.

Indian sub-continent having large cattle wealth can generate from dung and household wastes, huge amount of organic manures.

It is proven fact that organic wastes do poses fertility improvement nutrients. They had been valued even during modern advanced agricultural operations. Farmers and soil scientists must highlight the urgency of their application in raising food crops. Their utilisation must also be encouraged to negate the myth that chemical fertiliser's use can increase foodgrain production.

Soils are becoming less productive, mainly due to the continuous use of chemical fertilisers. They may become barren or desertic if they are used continuously. The United States, National Academy of Sciences and National Research Council has brought an elaborate report, whereby an appeal has been made to the government to encourage environmentally hazard free natural farming. Also to encourage farmers, world over, to adopt natural farming practices to produce healthy foodgrains, by discarding the use of toxic chemicals. In many countries including India, fertilisers are subsidised. Government policies encourage the use of fertilisers, overlooking their harmful effects, (over short and long term periods). Such environmentally conscious groups are also working in India to encourage the adoption of natural farming practices.

It is also a myth that more yield can be obtained by synthetic fertilisers. Farmers in United States have developed natural methods of cultivation. Farms in Knox country, Ohio have achieved 32 per cent and 40 per cent higher corn and soyabean yields by way of natural farming practices. Many farmers in Japan and more than 5 per cent farmers in United States are practising natural farming with higher profits and low investments.

Natural manures also help in improving the structure of the soils. Moisture holding capacity increases which helps maintain humus in soil. Crops can be grown even when water supplies are meagre. Natural farming will help in long run to improve soil cover, check depleting environment, control pollution and will also help in maintaining ecological balance. Various parts of India are semi-arid and run a high risk of becoming desertic, if not emancipated from scourge of chemical fertilisers. To save mankind natural farming system must get a priority. Because with increased literacy and awareness people are health conscious and like to have disease risk free food products from hazard free environment.

Natural farming practices need more care for crops, thus needing more time to be devoted in raising crop culture, which in turn will help in more output and slashing input costs considerably, thereby promoting economic well being and better status of the farming community. Because soil properties and qualities are basis for sustainable farming, which can be achieved by encouraging natural farming system.

TABLE 19.5: Average nutrient content of principal concentrated organic manures

Name of Organic Manure	Per cent Composition N	P_2O5	K_2O
Non-Edible Oil Cakes:			
Cotton Seed Cake	3.9	1.8	1.6
Mahuwa Cake	2.5	0.8	1.8
Neem Cake	5.2	1.0	1.4
Safflower Cake	4.9	1.4	1.2
Edible Oil Cake:			
Coconut cake	3.0	1.9	1.8
Cotton seed cake	6.4	2.9	2.2
Groundnut cake	7.3	1.5	1.3
Linseed cake	4.9	1.4	1.3
Rape-seed cake	5.2	1.8	1.2
Others			
Blood Meal	10.0	1.5	1.0
Fish Meal	4.10	3.9	0.3-1.5
Hoof & Horn	10.15	1.0	—
Raw Bone Meal	3.4	20.0	—
Poultry (waste, excreta) on dry wt. Basis	1.46	1.17	0.62

Source: Compiled from various sources.

TABLE 19.6: Agricultural inputs and production 1987-88

State	Agricultural Machinery in use in per cent	Irrigation by pumpsets (Elec+Diesel)	Fertilisers '000 tonnes	Food Production thousand Metric Tonnes	Area under Food grain thousand Hectares
U.P.	21.23	6,52,000	1,503	2,80,953	19,291
Punjab	27.97	4,52,000	1,112	1,70,654	5,288
Haryana	7.03	1,93,000	394	6,301.9	3,167
All India	100	47,97,000	8,696	1,38,413.9	1,18,966

Source: Complied from various sources.

REFERENCES

Acharya, C. N., 1954 : Compost', Your Phosphate Fertilisers, *Indian Farming*, Vol. 4(8).

Adhikari, G., 1989 : 'Call to Encourage Natural Farming', *Times of India.*

Cooke, W. G., 1969 : *'Fertilisers and Profitable Farming'* Crosby Wokwood & Sons Ltd., London.

Edward, C. A. and Health, G. W., 1963 : 'Role of Soil Animals in Breakdown of leaf Material', in J. Doeksen, and N. Vander Drift, (Eds) *Soil Organisms*, N. Holland Publishing, Amsterdam.

Idnani, M. A., and Walunjkar, W. G., 1954 : 'Utilisation of Some Nitrogenous Waste Materials an Manures', *Indian Journal of Agricultural Science*, Vol. 24.

Khan, A. R., 1952 : "Green Manuring For Better Crop Production", *Indian Farming*, Vol. 2(3).

Max. F. Millikan, 1969 : "The Dilemma of Agriculture in Underdeveloped countries", in David Hapgood (ed.), *No Easy Harvest*, Scientific Book Agency, Calcutta.

Parkinson, D., 1978 : 'Restoration of Soil Productivity', in M. W. Holdgate and M. J. Woodman, (eds) : *Breakdown and Restoration of Ecosystem*, Plenum Press, New York.

Sacheti, A. K., 1985 : *'Fertilisers and Manures'*, NCERT, New Delhi.

INDEX